PULSE GAS-DISCHARGE ATOMIC AND MOLECULAR LASERS

IMPUL'SNYE GAZORAZRYADNYE LAZERY NA PEREKHODAKH ATOMOV I MOLEKUL

ИМПУЛЬСНЫЕ ГАЗОРАЗРЯДНЫЕ ЛАЗЕРЫ НА ПЕРЕХОДАХ АТОМОВ И МОЛЕКУЛ

The Lebedev Physics Institute Series

Editors: Academicians D. V. Skobel'tsyn and N. G. Basov

P. N. Lebedev Physics Institute, Academy of Sciences of the USSR

Recent Volumes in this Series

Volume 35	Electronic and Vibrational Spectra of Molecules
Volume 36	Photodisintegration of Nuclei in the Giant Resonance Region
Volume 37	Electrical and Optical Properties of Semiconductors
Volume 38	Wideband Cruciform Radio Telescope Research
Volume 39	Optical Studies in Liquids and Solids
Volume 40	Experimental Physics: Methods and Apparatus
Volume 41	The Nucleon Compton Effect at Low and Medium Energies
Volume 42	Electronics in Experimental Physics
Volume 43	Nonlinear Optics
Volume 44	Nuclear Physics and Interaction of Particles with Matter
Volume 45	Programming and Computer Techniques in Experimental Physics
Volume 46	Cosmic Rays and Nuclear Interactions at High Energies
Volume 47	Radio Astronomy: Instruments and Observations
Volume 48	Surface Properties of Semiconductors and Dynamics of Ionic Crystals
Volume 49	Quantum Electronics and Paramagnetic Resonance
Volume 50	Electroluminescence
Volume 51	Physics of Atomic Collisions
Volume 52	Quantum Electronics in Lasers and Masers, Part 2
Volume 53	Studies in Nuclear Physics
Volume 54	Photomesic and Photonuclear Reactions and Investigation Method with Synchrotrons
Volume 55	Optical Properties of Metals and Intermolecular Interactions
Volume 56	Physical Processes in Lasers
Volume 57	Theory of Interaction of Elementary Particles at High Energies
Volume 58	Investigations in Nonlinear Optics and Hyperacoustics
Volume 59	Luminescence and Nonlinear Optics
Volume 60	Spectroscopy of Laser Crystals with Ionic Structure
Volume 61	Theory of Plasmas
Volume 62	Methods in Stellar Atmosphere and Interplanetary Plasma Research
Volume 63	Nuclear Reactions and Interaction of Neutrons and Matter
Volume 64	Primary Cosmic Radiation
Volume 65	Stellarators
Volume 66	Theory of Collective Particle Acceleration and Relativistic Electron Beam Emission
Volume 67	Physical Investigations in Strong Magnetic Fields
Volume 68	Radiative Recombination in Semiconducting Crystals
Volume 69	Nuclear Reactions and Charged-Particle Accelerators
Volume 70	Group-Theoretical Methods in Physics
Volume 71	Photonuclear and Photomesic Processes
Volume 72	Physical Acoustics and Optics: Molecular Scattering of Light; Propagation of Hypersound; Metal Optics
Volume 73	Microwave–Plasma Interactions
Volume 74	Neutral Current Sheets in Plasmas
Volume 75	Optical Properties of Semiconductors
Volume 76	Lasers and Their Applications
Volume 78	Research in Molecular Laser Plasmas
Volume 77	Radio, Submillimeter, and X-Ray Telescopes
Volume 79	Luminescence Centers in Crystals
Volume 80	Synchrotron Radiation
Volume 81	Pulse Gas-Discharge Atomic and Molecular Lasers

Proceedings (Trudy) of the P. N. Lebedev Physics Institute

Volume 81

PULSE GAS-DISCHARGE ATOMIC AND MOLECULAR LASERS

Edited by

N. G. Basov

P. N. Lebedev Physics Institute
Academy of Sciences of the USSR
Moscow, USSR

Translated from Russian by
Albin Tybulewicz

Editor, *Soviet Journal of Quantum Electronics*

CONSULTANTS BUREAU
NEW YORK AND LONDON

Library of Congress Cataloging in Publication Data

Main entry under title:

Pulse gas-discharge atomic and molecular lasers.

(Proceedings (Trudy) of the P. N. Lebedev Physics Institute; v. 81)
Translation of Impul'snye gazorazriàdnye lazery na perekhodakh atomov i molekul.
Bibliography: p.
1. Gas lasers. I. Basov, Nikolaĭ Gennadievich, 1922-   II. Series. Akademiia
nauk SSSR. Fizicheskiĭ institut. Proceedings; v. 81.
QC1.A4114 vol. 81 [TA1695] 530'.08s
                          [621.36'63]   76-57191
ISBN 978-1-4684-1628-2   ISBN 978-1-4684-1626-8 (eBook)
DOI 10.1007/978-1-4684-1626-8

The original Russian text was published by Nauka Press in Moscow in 1975 for the
Academy of Sciences of the USSR as Volume 81 of the Proceedings of the P. N. Lebedev
Physics Institute. This translation is published under an agreement with the Copyright
Agency of the USSR (VAAP).

PREFACE

This volume contains the results of many years of investigations of pulse gas-discharge lasers carried out at the Optical Laboratory of the Lebedev Physics Institute in Moscow. The two papers report mainly experimental results obtained in studies of pulse lasers utilizing translations in metals (Isaev and Petrash) and electronic transitions in diatomic molecules (Kaslin and Petrash). Population inversion mechanisms and the principal properties of the lasers are considered.

CONTENTS

Investigation of Pulse Gas-Discharge Lasers
  Utilizing Atomic Transitions
    A. A. Isaev and G. G. Petrash

Introduction. 1
  §1. Efficiency in a Gas-Discharge Laser 1
  §2. Review of the Literature 4

Chapter I. Kinetics of Output Power Saturation in a Three-Level System. 6
  §1. Level Populations and Saturated Power in a Three-Level System 7
  §2. Absence of Decay of Laser Levels. 8
  §3. Decay of Laser Levels by Emission of Spontaneous Radiation. 10
  §4. Decay of Laser Levels due to Collisions with Electrons 12
  §5. Pulse Superradiance due to Transitions from Resonance to Metastable
        Levels 14
  §6. Saturated Superradiance Power of Green Thallium Line 15

Chapter II. Apparatus and Measurement Techniques 18
  §1. Gas-Discharge Tubes and System for Evacuation and Filling with Gases 18
  §2. Resonators, Mirrors, Superradiance 20
  §3. Pulse Power Supply System 21
  §4. Measurement of Spectral and Time Characteristics 23
  §5. Measurement of Electrical Characteristics of Discharges 25

Chapter III. Pulse Laser Emission and Superradiance due to Transitions from
  Resonance to Metastable Levels in Atoms and Ions. 26
  §1. Superradiance due to Transitions in Helium Atoms. 26
  §2. Superradiance of Thallium Vapor 31
  §3. Pulse Superradiance due to Green Line of Thallium in TlI Vapor 43
  §4. Laser Emission and Superradiance due to Transitions in Lead. 49
  §5. Superradiance in Mercury Vapor. 58

Chapter IV. Pulse Laser Emission and Superradiance due to 2p−1s Transitions
  in Inert Gases 60
  §1. Superradiance and Laser Emission due to Transitions in Neon 61
  §2. Superradiance and Laser Emission due to Transitions in Argon, Krypton,
        and Xenon 73

Concluding Remarks 78
  §1. Pulse Lasers Utilizing Transitions from Resonance to Metastable Atomic
        Levels 78
  §2. Stimulated Emission due to Transitions in Ions 82

§3. Pulse Laser Emission and Superradiance due to 2p−1s Transitions in
     Inert Gases 83
§4. Conclusions 84

Literature Cited 85

Pulse Gas Lasers Utilizing Electronic Transitions
  in Diatomic Molecules
     V. M. Kaslin and G. G. Petrash

Introduction 89
§1. Characteristics of Lasers Utilizing Electronic Transitions in Molecules 89
§2. Review of the Literature and Formulation of the Problem 91
     List of Symbols 95

Chapter I. Gain of Electronic Transitions in Diatomic Molecules 96
§1. General Formula for Gain of Electronic-Vibrational-Rotational
     Transitions 96
§2. Distribution of Gain in Rotational Structure of Molecular Bands 99
§3. Dependence of Gain on Gas Temperature 104
§4. Calculation of Gain for 0−0 Band in Second Positive System of Nitrogen 105

Chapter II. Experimental Method 110
§1. Discharge Excitation System 110
§2. Measurement of Parameters of Excitation Pulses 112
§3. Spectral Measurement. Spectroscopic Apparatus 112
§4. Determination of Time Characteristics of Spontaneous and Stimulated
     Radiation 114
§5. Gas-Discharge Tubes. Vacuum System 114
§6. Systems for Cooling Working Gases 115
§7. Mirrors and Resonators. Superradiance and Laser Emission 116

Chapter III. Laser Emission and Superradiance in Second Positive
(Ultraviolet) System of Nitrogen Bands 117
§1. Characteristics of Laser Emission 117
§2. Influence of the Gas Temperature on Laser Emissions 119
§3. Laser Emission Spectra 120
§4. Discussion of Results 126

Chapter IV. Laser Emission and Superradiance in First Positive (Infrared)
System of Nitrogen Bands 135
§1. Characteristics of Laser Emission from Cooled Active Gas 135
§2. Laser Emission and Superradiance Spectra. Reversal of Intensity
     Alternation 137
§3. Discussion of Results 142

Chapter V. Laser Emission in Ångstrom (Visible) Band System of Carbon
Monoxide 145
§1. Characteristics of Laser Emission. Influence of Cooling 145
§2. Laser Emission Spectra 154
§3. Discussion of Results 158

Conclusions 163

Appendix. Summary Table of All Known Laser Emission Lines due to Electronic-
    Vibrational-Rotational Transitions in Molecules Arranged in Order of
    Increasing Wavelength.   170

Literature Cited.   184

INVESTIGATION OF PULSE GAS-DISCHARGE LASERS
UTILIZING ATOMIC TRANSITIONS

A. A. Isaev and G. G. Petrash

An investigation was made of pulse gas-discharge lasers utilizing transitions from resonance to metastable levels in helium, thallium, and lead atoms and in mercury ions. All these lasers had high ultimate efficiencies. Population inversion mechanism of the green thallium line emitted from thallium iodide vapor was studied. Investigations were made of the pulse laser emission and superradiance due to 2p-1s transitions in neon, argon, krypton, and xenon; the population inversion mechanism responsible for these transitions was confirmed. The experimental data obtained in the present study and the results of other workers were used to analyze the prospects of pulse lasers of the type considered. The active media and transitions, which should ensure pulse stimulated emission, were identified.

INTRODUCTION

One of the principal aims in the development of gas lasers is to construct systems characterized by a high efficiency and a high specific output power. This is particularly important in the case of short-wavelength lasers emitting in the ultraviolet, visible, and near infrared range. The low efficiency of the gas lasers emitting short-wavelength radiation reduces considerably their potential applications in science and technology. The aim of the investigation reported below was to identify the factors limiting the efficiency and output power of pulse gas-discharge lasers and to find ways of increasing these parameters. We shall show that pulse gas lasers are promising in these respects.

§1. Efficiency in a Gas-Discharge Laser

It is convenient to start by considering the principal characteristics of an active medium, which govern the ultimate efficiency of conversion of excitation energy into stimulated radiation. We shall be interested only in systems with a sufficiently high efficiency. Therefore, we shall assume that the rate of direct population of the lower active (laser) level from the ground state is considerably slower than the rate of population of the upper level. Then, the quantum efficiency of transitions in a cw laser can be expressed in the form

$$\eta = f_p \frac{h\nu_{st}}{E_u},$$

where $h\nu_{st}$ is the energy of a photon emitted as a result of a stimulated transition, and E_u is the energy of the upper laser level. The ratio $h\nu_{st}/E_u$ shows what proportion of the upper laser level energy is utilized in stimulated emission. The factor f_p is the pumping efficiency

1

and it represents that fraction of the excitation energy supplied to the active medium, which is converted to excitation of the upper laser level.

Steady-state population inversion for any $i \rightarrow k$ transition can be maintained if, in addition to selective pumping of the upper laser level, the probability of the laser transition A_{ik} is less than the total probability of decay of the lower laser level A_k. This requirement imposes serious restrictions on the selection of the transitions suitable for steady-state inversion. In the majority of gas lasers emitting in the visible, ultraviolet, and near infrared range the decay (deactivation) of the lower laser level is due to the emission of spontaneous radiation. An analysis report in [1, 2] shows that the decay of the lower laser level often ensures the necessary rate of relaxation but it reduces the output power and efficiency because the transition must then take place between relatively high levels. For a typical cw gas laser the energy of the upper level E_u is of the order of 15-30 eV and the ratio $h\nu_{st} / E_u$ rarely exceeds 0.1. The use of high levels has also an unfavorable influence on the value of f_p. In a typical gas-discharge plasma the energy is lost mainly due to the excitation of the lowest levels and ionization. Only about 1% of the total energy is used to excite higher levels. Consequently, the efficiency of cw gas-discharge lasers is $\eta \sim 10^{-3}$-10^{-4}. A possible way of overcoming these difficulties is represented by the idea of a collision laser in which the lower level is deactivated by collisions with heavy particles [1, 2]. However, this idea has been put into practice only in the middle range of infrared frequencies utilizing transitions between low-lying vibrational levels but it has proved that lasers with a high efficiency and a high output power can be obtained. The best example of such a system is the well-known CO_2 laser [3, 4].

Relaxation of the lower laser level is unnecessary in the case of pulse systems emitting radiation at the beginning of the excitation pulse. This extends considerably the choice of the laser transitions and should make it possible to achieve stimulated emission in a wide range of wavelengths and with better characteristics than those of cw lasers; this applies particularly to the efficiency and peak output power.

The present paper reports an investigation of the population inversion in atomic systems (atoms and atomic ions) so that we shall confine our attention to transitions in atoms and ions. In atomic systems the major part of the discharge energy is usually converted to the excitation of the first resonance level. For example, about 60% of the total energy supplied to a steady-state discharge in mercury is used to excite the resonance level [5, 6]. The first resonance level usually has the largest electron-excitation cross section. Therefore, it is desirable to use the first resonance level as the upper laser level. The lower level in atomic systems can only be a metastable level located below the first resonance level. This use of the metastable level should ensure efficient population inversion because the electron-excitation cross sections of forbidden transitions are usually smaller than those of allowed transitions. Thus, we may expect that in a typical case a metastable level is populated from a resonance level and transitions from the resonance to the metastable level should establish a population inversion at the beginning of an excitation pulse.

Under normal conditions the lifetime of metastable levels is relatively long. Therefore, a population inversion in transitions of this kind can usually be achieved only at the beginning of an excitation pulse. In principle, the deactivation of the metastable level can occur quite rapidly because of collisions with heavy particles (collision laser), but in practice it is difficult to achieve the relaxation rate necessary for steady-state inversion. However, the lower level can be deactivated by collisions not during the excitation of the upper level but in the interval between the excitation pulses. In this case the rate of relaxation of the lower level governs the permissible repetition frequency of the output pulses and the average output power. Such lasers are known as cyclic [7]. They can be regarded as an intermediate stage in the development of highly efficient cw collision lasers [8].

Pulse laser emission due to transitions terminating at metastable levels is frequently called self-terminating because its duration is limited by the properties of the transition itself. The expression for the efficiency of such lasers is somewhat different because laser emission continues until the populations of the laser levels become equal. Part of the population of the upper laser level remains unused: this part depends on the ratio of the statistical weights of the upper g_u and the lower g_l levels. Allowance for this factor gives

$$\eta = f \frac{h\nu_{st}}{E_u} \frac{g_l}{g_l + g_u} = f\eta_{ult}.$$

We shall call η_{ult} the ultimate efficiency of the transition, i.e., the efficiency attained when all the excitation energy is used to excite the upper laser level and the pumping efficiency is $f = 1$. The factor $g_l/(g_l + g_u)$ is of the order of unity: it is usually between $^2/_3$ and $^1/_3$. However, the factor $h\nu_{st}/E_u$ may be considerably greater than in the steady-state (continuous emission) case. For atoms with a relatively low-lying lower level this factor may reach 0.5-0.7. The ultimate efficiency η_{ult} can be calculated if the laser transition is known. On the other hand, an estimate of f is always difficult to obtain because this factor depends strongly on the properties of the active medium and the experimental conditions. For resonance levels the factor f may be of the order of 0.5. Multiplying out all these factors, we obtain $\eta \approx 25\%$. Such an efficiency would be a considerable achievement. However, it should be noted that the realization of high efficiencies under pulse conditions requires that several additional conditions be satisified. First of all, a pulse population inversion exists for a limited and sometimes very short time. Therefore, a high efficiency is obtained if the duration of the excitation pulses is of the order of the population inversion lifetime. It the excitation pulse is longer than the inversion lifetime, not only is the energy lost in a useless manner, but also the conditions for the next stimulanted emission pulse may deteriorate. Consequently, studies of the time characteristics of the stimulated radiation pulses become extremely important, i.e., it is necessary to determine the position of stimulated emission pulses relative to the excitation pulses, as well as their duration and shape. Equally important are the studies of the duration and shape of the excitation pulses.

Moreover, we must bear in mind that, in contrast to continuous discharges, a considerable energy is lost in pulse discharges to establish a plasma state. Rough estimates indicate that the energy needed for the ionization and heating of electrons in a plasma may be comparable with the energy lost in the excitation of the active levels. When these points are allowed for and the difficulty of finding atoms with an ideal energy level scheme for population inversion is considered, it is found that the real efficiency due to transitions from resonance to metastable levels may be only ~10%. However, this is still a relatively high efficiency for pulse conditions and it is desirable to study transitions from resonance to metastable levels.

It is interesting to estimate also other characteristics of pulse lasers, particularly their peak output power. We shall only obtain rough estimates of this power on the basis of very general considerations. The next chapter will be concerned with detailed calculations of the stimulated radiation power.

If the laser threshold is exceeded significantly, the output power is governed primarily by the rate of pumping of the upper laser level. We shall estimate the attainable pumping rate on the assumption that the excitation is due to direct electron impact from the ground state. The pumping rate is then given by the expression

$$Q = N_0 \langle \sigma v_e \rangle n_e,$$

where N_0 is the population of the ground state, $\langle \sigma v_e \rangle$ is the average (over the electron velocities) product of the excitation cross section of the level in question and the electron velocity, and n_e is the electron density. Under favorable conditions the excitation cross section may be of the order of 10^{-16} cm^2. Estimates of cross sections for real transitions give lower values. The electron velocity can be assumed to be quite close to 10^8 cm/sec. The electron density in pulse discharges may reach 10^{17} cm^{-3} but at very high electron densities we have to allow also for the deactivation of levels due to collisions with electrons so that the population of the upper level may cease to rise with increasing electron density. An acceptable electron density is then $n_e = 10^{15}$ cm^{-3}. For these parameters the pumping rate is $10^7 N_0$ excitation events in 1 cm^3 per 1 sec. Clearly, it is desirable to use the highest possible densities of the active particles. However, in practice an increase in the working density meets with certain difficulties. For example, if we assume that $N_0 = 10^{17}$ cm^{-3}, which corresponds to a pressure of ~ 3 Torr at room temperature, i.e., if we postulate a density close to real densities in existing lasers, we find that the pumping rate is $Q \approx 10^{24}$ cm$^{-3} \cdot$ sec^{-1}. Calculations indicate that the saturated output power of pulse systems, expressed in the number of transitions in 1 cm^3 per 1 sec laser emission maximum, is slightly lower (because of the population of the lower level) but still of the same order as the pumping rate of the upper level. If we assume that the laser emission occurs in the visible part of the spectrum, it follows from the above estimates that the specific saturated output power should be $P = 10^5$ W/cm^3. This estimate is only accurate to within an order of magnitude but it shows that high peak output powers are possible. For example, an active medium of 1000 cm^3 volume may emit a peak output power of 100 MW.

A high specific output power is obtained for laser transitions in which the upper level is excited efficiently by electrons. Thus, the choice of a transition for the highest output power agrees with the choice of the transition for the highest efficiency. Essentially, attempts to increase the efficiency and the specific power are simply two aspects of the same problem.

Thus, pulse gas-discharge lasers should, in principle, be characterized by high efficiencies and high peak output powers. Estimates also suggest that we can expect laser emission in a wide range of wavelengths and a high pulse repetition frequency. The practical realization of these considerable potential capabilities will depend on the success in the search for a medium with a near-ideal structure of the levels and the way the practical difficulties in the construction of such lasers are overcome.

We have shown above that high output powers and high efficiencies should be obtained using transitions from resonance to metastable levels in atoms and atomic ions. Consequently, we searched for and studied transitions of this kind for use in pulse lasers. Moreover, we investigated lasers utilizing transitions of the 2p-1s type in inert gas atoms. The population inversion mechanism in the latter lasers was not clear but it has been suggested that it is similar to the inversion involving resonance and metastable levels. Additional advantages of the inert gas lasers are their simplicity and the convenience of working with inert gases. Consequently, they were included in our investigation.

§ 2. Review of the Literature

When the present investigation was started, the published information on pulse laser emission due to transitions from the first resonance levels to metastable levels was available only for lead, copper, and manganese atoms and also for calcium ions.

Stimulated transitions of this type were first reported in 1965 for lead vapor [9]. The stimulated radiation was due to the $6p7s\ ^3P_1^0 - 6p^2\ ^1D_2$ ($\lambda = 7229$ Å) transition. A high gain was achieved in the first study of this transition and a clearly observed superradiance was reported.

Somewhat later stimulated radiation was obtained as a result of transitions in manganese atoms [10]. It was observed for five transitions of the $y^6P_J^0 - a^6D_J$ type in near infrared. Once more, a high gain was reported. The peak output power reached 300 W. The duration of stimulated emission in the green part of the spectrum was about 20 nsec.

The discovery of laser emission due to transitions in manganese was soon followed by a report of superradiance due to transitions in copper atoms [11]: the transitions in question were $4s\ ^2P^0_{3/2} - 4s^2\ ^2D_{5/2}$ ($\lambda = 5105$ Å) and $4s\ ^2P^0_{1/2} - 4s^2\ ^2D_{3/2}$ ($\lambda = 5782$ Å). The peak output power was 1.2 kW and the pulse duration was 20 nsec.

Pulse laser emission and superradiance were also reported in [7] for two transitions in calcium ions: $4p\ ^2P^0_{3/2} - 3d^2D_{5/2}$ ($\lambda = 8541$ Å) and $4p\ ^2P^0_{1/2} - 3d^2D_{3/2}$ ($\lambda = 8662$ Å). The peak output power due to these transitions was 30 W and the pulse duration was about 30 nsec. A brief review of pulse lasers utilizing transitions from resonance to metastable levels known at the time was given in [7]. Five conditions which efficient pulse gas-discharge lasers should satisfy were formulated there. These conditions were used as a basis of a list of transitions for which pulse stimulated emission could be expected.[7]. All the available experimental results confirmed that population inversion was due to direct electron excitation of resonance levels and weak excitation of metastable levels by the leading edges of excitation pulses.

A technique for the excitation of discharges in difficult-to-volatilize vapors at temperatures up to 1500°C was described in [12]. A discharge tube was made of aluminum oxide and its central part was enclosed in a special tubular heater. The electrodes and windows were outside the heated zone. Neon and helium were used as buffer gases.

Our investigation of pulse gas-discharge lasers started in 1966 with an analysis of conditions for achieving efficient laser emission as a result of atomic transitions. We analyzed all the atoms in the periodic system and selected the promising ones for an experimental study. The aim was that the working temperature of the active medium was not too high, preferably not higher than 1000°C. Serious difficulties which would create problems in practical use were encountered at higher temperatures. Among the pulse lasers mentioned so far the lead laser had a working temperature of about 1000°C, that utilizing manganese transitions operated at 1200°C, and the corresponding working temperatures of the copper and calcium lasers were 1500°C and 700°C, respectively.

Although laser emission as a result of transitions from resonance to metastable levels have been observed in several systems, it was not clear to what extent the analysis of the population inversion processes involved in these transitions was applicable to other atoms with similar level structures. In particular, it was not clear why only one transition was active among several possible transitions of similar kind. Therefore, it seemed desirable to increase the number of pulse gas lasers, so that more reliable conclusions about stimulated transitions in various atoms could be drawn.

Finally, it also seemed desirable to investigate atoms with a high ultimate efficiency of the transitions. When all these considerations were taken into account, it was decided to study helium, thallium, lead, mercury, bismuth, and barium. All these systems should operate below 1000°C. In some cases the ultimate efficiency should be very high. (Detailed characteristics of the investigated atoms and the justification for their selection will be given below.) We also carried out some experiments on molecules. These were related to our attempts to introduce atoms into a discharge at relatively low temperatures.

We shall now analyze the published information on the 2p-1s transitions in inert gas atoms which was available when we started the present investigation. Pulse laser emission and superradiance due to the 2p-1s transitions (Paschen notation) in neon were observed almost simultaneously in several independent studies [13-16]. Pulse laser emission due to similar transitions in krypton and xenon was reported soon after [17]. For reasons which were not clear, laser emission due to transitions in argon was not obtained.

All these atoms — neon, argon, krypton, and xenon — have a similar level structure, so that the population inversion mechanism should be the same. Transitions resulting in laser emission and superradiance terminate at metastable or quasimetastable levels. Thus, they are similar to the transitions discussed above. However, the upper laser levels in these transitions are not coupled by allowed optical transitions to the ground states of the atoms. Thus, the main condition − excitation of the upper level by a resonance transition − is not satisfied. Nevertheless, high-gain pulse superradiance was obtained as a result of these transitions and its properties were similar to those of the superradiance resulting from transitions between resonance and metastable levels in other atoms. The population inversion mechanism of the transitions in the inert atoms was not clear. Three different points of view were put forward in the literature (they are discussed in detail in Chap. IV). Stimulated emission as a result of these transitions would be of considerable practical interest. In contrast to the atoms for which laser emission was obtained as a result of transitions from resonance to metastable levels, inert gases would not require any heating and they would be more convenient to use than any other gases. This is because inert gases (even spectroscopically pure) are relatively easily available and their properties are well known from their use in the manufacture of gas-discharge devices. The gains of the transitions in question are very high. Finally, high peak output powers and very short pulses can be obtained (particularly from neon). All these factors stimulated out.interest in the superradiance due to 2p-1s transitions in neon, argon, krypton, and xenon.

A characteristic feature of the lasers investigated by us was a very high gain. In most cases stimulated emission was observed in systems without a resonator with just one mirror or without mirrors at all. Emission from an active volume highly extended in one direction (we used relatively long tubes) resembled an ordinary emission of a laser with a resonator. The divergence was governed by the geometric aperture of the active volume. This phenomenon is called by various names in the literature. We shall employ the term "superradiance" used also in our earlier papers. The term superluminescence is also used. We shall employ "laser emission" to describe the situation in which an active medium is in an optical resonator and "superradiance" for the case when there is no resonator and the radiation emitted from the active medium is observed without mirrors or with just one mirror.

Following the review of the published literature, we set ourselves the following tasks.

1. We intended to search for pulse laser emission or superradiance due to transitions from resonance levels in helium, thallium, mercury, and several other atoms, as well as to find the new transitions in lead. We also planned to observe laser emission or superradiance due to transitions in argon, or to account for their absence.

2. We intended to carry out a detailed experimental investigation of the properties of laser emission and superradiance due to these transitions. We were proposing to identify the population inversion mechanism and the factors governing the efficiency of laser emission and the attainable specific peak output power. Transitions with a high ultimate efficiency were to be given special attention.

Many papers appeared during our investigation. Those papers which were directly relevant to the problems we investigated will be discussed as necessary.

We shall conclude our paper with a brief chapter reviewing the available information on pulse gas-discharge lasers.

CHAPTER I

KINETICS OF OUTPUT POWER SATURATION IN A THREE-LEVEL SYSTEM

In this chapter we shall calculate theoretically the saturated output power of pulse stimulated radiation emitted by a three-level system. This system is a good approximation for lasers utilizing transitions from resonance to metastable levels because the excitation is confined mainly to the resonance level and the influence of the other levels is slight. This approximation is particularly good in the case of high-efficiency lasers.

The gain of the majority of lasers of interest to us is so high that saturation is reached relatively easily even in small-volume active media. Consequently, the results of a calculation of the saturated power should give a good description of the stimulated radiation power emitted by real lasers. The advantage of calculations dealing with the limiting case of total saturation is that they are relatively simple.

An approximate method for the calculation of saturated output power, based on the assumption of equality of the populations of the laser levels, is given in [18]. This method is used in [18, 19] in a calculation of the saturated output power of ultraviolet radiation emitted from molecular nitrogen. The results reported there show that the method gives a good description of the experimental dependences. However, the calculations in [18, 19] are numerical (carried out on a computer) and they apply to specific experimental conditions and definite characteristics of the laser levels.

We shall attempt to calculate the saturated power employing the same approximate method as in [18] but using a simplified model which can be solved analytically. An analytic solution is very desirable because it gives a clear idea on the nature of the processes occurring in a laser under investigation and on the influence of various parameters of a laser system so that it is possible to compare various systems from the point of view of obtaining stimulated radiation with specified characteristics.

§1. Level Populations and Saturated Power

in a Three-Level System

We shall consider a three-level system shown in Fig. 1 and we shall discuss the processes of electron excitation and deexcitation as well as deactivation due to emission of spontaneous radiation. Following [18], we shall calculate the stimulated radiation power P per unit volume. Equations for the populations of the upper (resonance) N_r and lower (metastable) N_m levels are

$$\frac{dN_r}{dt} = N_0 q_{0r} n_e + N_m q_{mr} n_e - N_r [A_r + (q_{r0} + q_{rm}) n_e] - P, \tag{1}$$

$$\frac{dN_m}{dt} = N_0 q_{0m} n_e + N_r (q_{rm} n_e + A_{rm}) - N_m [A_m + (q_{m0} + q_{mr}) n_e] + P. \tag{2}$$

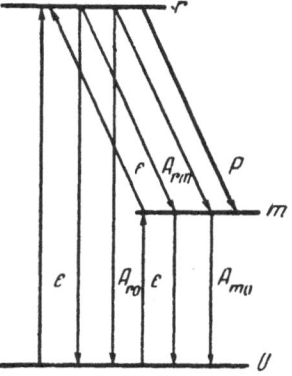

Fig. 1. Excitation and deexcitation processes in a three-level system.

We are assuming that $N_r \ll N_0$, $N_m \ll N_0$, $N_0 = $ const. Here, N is the population of level i; n is the electron density; $q_{ik} = \langle \sigma_{ik} v_e \rangle$ is the average (over the electron velocity distribution) product of the electron excitation (or deexcitation) cross section for the $i \to k$ transition; A_i is the total probability of the spontaneous decay (deactivation) of level i; P is the stimulated radiation power due to the $r \to m$ transition, expressed as the number of transitions per unit volume per unit time.

In calculating the saturated stimulated radiation power we shall assume [18] that the field is sufficiently strong for the population inversion to vanish but, in contrast to [18], we shall include statistical weights of the laser levels g_r and g_m and we shall assume that $N_r/g_r = N_m/g_m$, i.e., $N_r = \varepsilon N_m$, $\varepsilon = g_r/g_m$. In this case Eqs. (1) and (2) become

$$\frac{d}{dt}(N_m + N_r) = \frac{1+\varepsilon}{\varepsilon}\frac{dN_r}{dt} = N_0(q_{0r} + q_{0m})n_e - N_r\left[\left(A_r - A_{rm} + \frac{A_m}{\varepsilon}\right) + \left(q_{r0} + \frac{q_{m0}}{\varepsilon}\right)n_e\right], \qquad (3)$$

$$(1+\varepsilon)P = N_0(q_{0r} - \varepsilon q_{0m})n_e - N_r\left\{(A_r + \varepsilon A_{rm} - A_m) + n_e\left[q_{r0} - q_{m0} + (1+\varepsilon)q_{rm} - \frac{1+\varepsilon}{\varepsilon}q_{mr}\right]\right\}. \quad (4)$$

Equation (3) describes the observation that under saturation conditions the populations of the laser levels seem to be shared. The time dependence of the total population of the laser levels is governed by the sum of the rates of populations of the levels and the sum of the rates of their deactivation. Equation (4) describes the time dependence of the saturated stimulated radiation power, needed to maintain population inversion at zero level. Positive values of P correspond to stimulated radiation and negative values to absorption. Equation (4) is algebraic so that the possibility of obtaining an analytic solution is determined entirely by the solution of Eq. (3). In this equation q_{ik} and n_e are functions of time governed by specific experimental conditions. The nature of these functions determines the nature of the solution of Eq. (3). An analytic solution can be obtained only for relatively simple dependences. We shall consider a model which is governed by the particular selection of these dependences. We shall assume that the electron density increases linearly with time and that the quantities q_{ik} are independent of time: $n_e = \alpha t$, $q_{ik} = $ const. This model describes approximately the initial period of the discharge of a capacitance when the voltage across the capacitance is still close to its initial value and the rise of the current is limited by the circuit inductance. This situation is usually encountered in the excitation of steep-edged pulses. However, even for these very simple dependences the general solution of Eq. (3) contains expressions with error functions and is very cumbersome. Therefore, we shall have to confine ourselves to some limiting cases.

§ 2. Absence of Decay of Laser Levels

We shall first consider the earliest stages of an excitation pulse, i.e., we shall consider time intervals much shorter than the characteristic times of decay (deactivation) of the laser levels. We can then simplify Eq. (3) by dropping the term proportional to N_r and describing the decay of the laser levels. We then readily obtain

$$N_r = \frac{\alpha \varepsilon N_0}{1+\varepsilon}(q_{0r} + q_{0m})\frac{t^2}{2}. \qquad (5)$$

We can show that this approximation is valid subject to the conditions

$$t \ll \frac{3}{A_r - A_{rm} + A_m/\varepsilon}, \qquad t \ll \sqrt{\frac{4(1+\varepsilon)}{\alpha\varepsilon(q_{r0} + q_{m0}/\varepsilon)}}. \qquad (6)$$

The time dependence of the saturated power is then governed by the decay due to the emission of spontaneous radiation and due to collisions with electrons. The saturated power is then given by

$$P = \frac{\alpha N_0}{1+\varepsilon}(q_{0r} - \varepsilon q_{0m})\, t \left\{ 1 - \frac{\varepsilon}{2} \frac{q_{0r} + q_{0m}}{q_{0r} - \varepsilon q_{0m}} \left[A_{rm} t + \alpha t^2 \left(q_{rm} - \frac{q_{mr}}{\varepsilon} \right) \right] \right\}. \tag{7}$$

We shall assume that the transitions due to collisions with electrons can be ignored. Then,

$$P = \frac{\alpha N_0}{1+\varepsilon}(q_{0r} - \varepsilon q_{0m})\, t \left\{ 1 - \frac{\varepsilon}{2} \frac{q_{0r} + q_{0m}}{q_{0r} - \varepsilon q_{0m}} A_{rm} t \right\}. \tag{8}$$

We can easily see that in this case at the moment

$$t_{\max} = \frac{1}{\varepsilon A_{rm}} \frac{q_{0r} - \varepsilon q_{0m}}{\varepsilon (q_{0r} + q_{0m})} \tag{9}$$

the saturated power reaches its maximum

$$P_{\max} = \frac{1}{2} \frac{\alpha N_0}{1+\varepsilon} \frac{q_{0r} - \varepsilon q_{0m}}{\varepsilon A_{rm}} \frac{q_{0r} - \varepsilon q_{0m}}{q_{0r} + q_{0m}}. \tag{10}$$

The total duration of stimulated emission is thus $t_k = 2t_{\max}$ so that the peak output power rises linearly with the steepness of the rise of the electron density α and is inversely proportional to the probability of the laser transition. Under favorable conditions we can select a level system so that $q_{0m} \ll q_{0r}$. Then,

$$t_{\max} \approx \frac{1}{\varepsilon A_{rm}}, \qquad t_k \approx \frac{2}{\varepsilon A_{rm}}, \qquad P_{\max} \approx \frac{\alpha N_0 q_{0r}}{2\varepsilon (1+\varepsilon) A_{rm}}. \tag{10'}$$

It is clear from Eqs. (9), (10), and (10') that the quantity A_{rm} should be reduced as much as possible in order to achieve the highest peak output power and duration of stimulated radiation. However, if A_{rm} is reduced, we find that beginning from a certain value the decay due to the laser transition becomes governed by electron collisions. If the spontaneous transition probability A_{rm} is ignored, we find that

$$P = \frac{\alpha N_0}{1+\varepsilon}(q_{0r} - \varepsilon q_{0m})\, t \left\{ 1 - \frac{\alpha \varepsilon}{2} \frac{q_{0r} + q_{0m}}{q_{0r} - \varepsilon q_{0m}} \left(q_{rm} - \frac{q_{mr}}{\varepsilon} \right) t^2 \right\}. \tag{11}$$

In this case, we have

$$t_{\max}^2 = \frac{1}{3\beta}, \qquad t_k^2 = \frac{1}{\beta}, \quad \text{where } \beta = \frac{\alpha \varepsilon}{2} \frac{q_{0r} + q_{0m}}{q_{0r} - \varepsilon q_{0m}} \left(q_{rm} - \frac{q_{mr}}{\varepsilon} \right),$$

$$P_m = \frac{\alpha N_0}{1+\varepsilon}(q_{0r} - \varepsilon q_{0m}) \frac{2}{3} \sqrt{\frac{1}{3\beta}} \sim \sqrt{\alpha}. \tag{12}$$

If $q_{0m} \ll q_{0r}$ and $\beta \simeq \alpha \varepsilon / 2 \, (q_{rm} - q_{mr}/\varepsilon)$, we find that

$$P = \frac{2}{3} \frac{\sqrt{\alpha} N_0}{1+\varepsilon} q_{0r} \sqrt{\frac{2}{3\varepsilon (q_{rm} - q_{mr}/\varepsilon)}}. \tag{12'}$$

In this limiting case the peak power is proportional to $\sqrt{\alpha}$. Combining the expression for the duration of stimulated emission t_k and the conditions of validity of the adopted approximation (6), we obtain a new condition:

$$A_{rm} \gg \frac{2}{3\varepsilon} \left(A_r - A_{rm} + \frac{A_m}{\varepsilon} \right), \tag{13}$$

$$\left(q_{rm} - \frac{q_{mr}}{\varepsilon} \right) \gg \frac{q_{r0} + q_{m0}/\varepsilon}{2(1+\varepsilon)}, \tag{13'}$$

which shows that the adopted approximation is justified if the probabilities of the spontaneous or collision-induced transitions between the laser levels are much higher than the sum of the probabilities of the decay of the laser levels by all mechanisms except the laser action.

It is clear from general considerations and from the above formulas that a high peak output power and a long duration of stimulated emission can be ensured by selecting such laser levels that the probabilities of spontaneous and collision-induced transitions between them are relatively low.* They may be of the order of the probabilities of the decay of the laser levels by other mechanisms and then our approximation is no longer valid. In view of this, it is necessary to allow for the influence of the decay of the laser levels by mechanisms other than the laser action.

§3. Decay of Laser Levels by Emission
of Spontaneous Radiation

We shall begin our analysis from the case of low electron densities. We shall assume that in a time interval of interest to use we have

$$n_e = at \ll \frac{A_r - A_{rm} + A_m/\varepsilon}{q_{r0} + q_{m0}/\varepsilon}.$$

(14)

This condition means that the decay of the laser levels to the ground state is due to the emission of spontaneous radiation. Then, Eq. (3) is easily solved and we can obtain an expression for N_r and for the saturated power in the form

$$N_r = \frac{N_0 a \varepsilon}{1 + \varepsilon} \frac{q_{0r} + q_{0m}}{\rho^2} (x - 1 + e^{-x}),$$
$$P = M \{x - (\theta + \varkappa x)[x - 1 + e^{-x}]\}.$$

(15)

Here,

$$x = \rho t, \qquad \rho = \frac{\varepsilon}{1 + \varepsilon}(A_r - A_{rm} + A_m/\varepsilon), \qquad M = N_0 a \frac{q_{0r} - \varepsilon q_{0m}}{\varepsilon(A_r - A_{rm}) + A_m},$$
$$\varkappa = \frac{q_{0r} + q_{0m}}{q_{0r} - q_{0m}\varepsilon} \frac{\alpha(1 + \varepsilon)}{\varepsilon} \frac{(1 + \varepsilon)q_{rm} - (1 + \varepsilon)q_{mr}/\varepsilon}{(A_r - A_{rm} + A_m/\varepsilon)^2},$$
$$\theta = \frac{q_{0r} + q_{0m}}{q_{0r} - \varepsilon q_{0m}} \frac{A_r + \varepsilon A_{rm} - A_m}{A_r - A_{rm} + A_m/\varepsilon}.$$

We have ignored here the deexcitation of the laser levels by electron impact to the ground state. However, we may find that we have to include the exchange of energy between the laser levels as a result of collisions with electrons. Consequently, the expression for the saturated power in the system (15) includes a term $\varkappa x$ which allows for the exchange of energy between the laser levels due to collisions with electrons.

We shall first consider electron densities so low that this exchange can be ignored. Then, the solution becomes

$$P = M[x - \theta(x - 1 + e^{-x})].$$

(16)

Figure 2a gives the dependences $P(x)/M$ for different values of θ. We can see that if $\theta > 1$, the dependence $P(x)$ has a maximum. We can easily show that the moment at which this

---

* However, it should be noted that if the probability of transitions between the laser levels is low, it is difficult to achieve superradiance and saturation conditions.

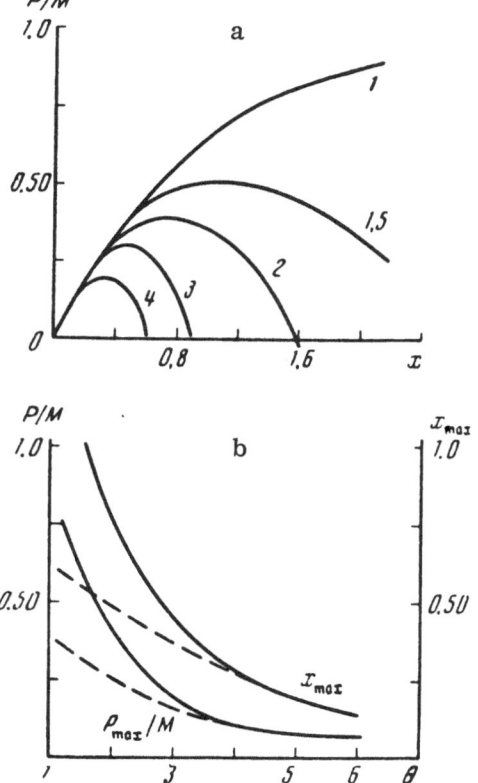

Fig. 2. Saturated power in the case of decay of laser levels due to emission of spontaneous radiation: a) time dependences of the saturated power for a low electron density, plotted for different values of the parameter θ; b) dependences of maximum saturated power P_{max} and of the time taken to reach this maximum x_{max} on the parameter θ. The continuous curves are obtained ignoring exchange between the laser levels due to collisions with electrons; the dashed curves are obtained allowing for such exchange.

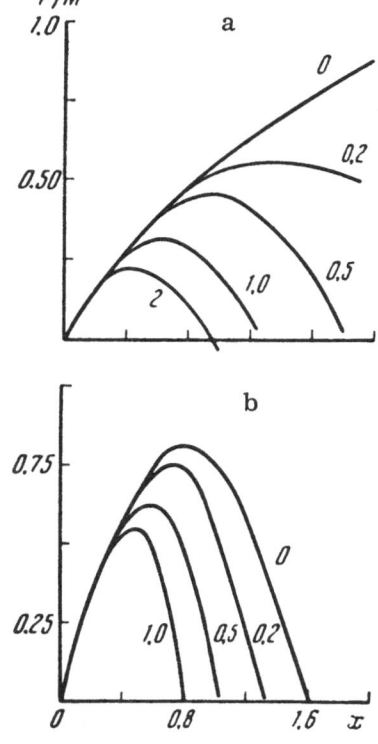

Fig. 3. Time dependences of the saturated power obtained allowing for collisions with electrons. The numbers alongside the curves are the values of $\varkappa$.

maximum is reached and the corresponding maximum saturated power are then described by

$$x_{max} = \ln \frac{\theta}{\theta - 1}, \qquad t_{max} = \frac{1}{\rho} \ln \frac{\theta}{\theta - 1},$$

$$P_{max} = M \left[1 - (\theta - 1) \ln \frac{\theta}{\theta - 1} \right]. \qquad (17)$$

The dependences $x_{max}(\theta)$ and $P_{max}(\theta)/M$ are plotted in Fig. 2b.

We shall now see the consequences of allowance for the exchange of energy between the laser levels due to collisions with electrons. In this case we can no longer drop the term $\varkappa x$ from Eq. (15). We can easily see that inclusion of this level reduces the saturated power and the reduction increases with x. This is illustrated in Fig. 3, which shows the dependence $P(x)/M$ for $\theta = 1$ (Fig. 3a) and for $\theta = 2$ (Fig. 3b), and for different values of the parameter $\varkappa$. It is clear from Fig. 3 that the higher the value of $\varkappa$, the shorter is the time during which the power rises, so that the peak output power and the duration of stimulated emission pulses both decrease. Figure 2b shows, by way of example, the dependences $x_{max}(\theta)$ and $P_{max}(\theta)/M$ which are represented by dashed curves. We can see that at high values of θ the influence of the term $\varkappa x$ is weak.

§ 4. Decay of Laser Levels due to Collisions

with Electrons

We shall now consider the other extreme case. We shall assume that during a time interval of interest to us (with the exception of the very first moments), we have

$$n_e = \alpha t \gg \frac{A_r - A_{rm} + A_m/\varepsilon}{q_{r0} + q_{m0}/\varepsilon}. \qquad (18)$$

This condition is opposite to that in Eq. (14) and it means that we can ignore the radiative decay of the laser levels to the ground state and consider only the decay due to collisions with electrons. In this case, Eq. (3) has the solution

$$N_r = N_0 \frac{q_{0r} + q_{0m}}{q_{r0} + q_{m0}/\varepsilon} (1 - e^{-y^2}), \qquad (19)$$

whereas the solution for the saturated power is

$$P = R [y - (\psi + ky)(1 - e^{-y^2})],$$

$$\text{where} \quad y = \gamma t, \quad \gamma^2 = \frac{\alpha \varepsilon}{2(1 + \varepsilon)} \left(q_{r0} + \frac{q_{m0}}{\varepsilon} \right),$$

$$R = \frac{\alpha N_0}{\gamma (1 + \varepsilon)} (q_{0r} - \varepsilon q_{0m}), \qquad \psi = \frac{\gamma}{\alpha} \frac{(q_{0r} + q_{0m})(A_r + \varepsilon A_{rm} - A_m)}{\left(q_{r0} + \frac{q_{m0}}{\varepsilon} \right)(q_{0r} - \varepsilon q_{0m})}, \qquad (20)$$

$$k = \frac{(q_{0r} + q_{0m})[q_{r0} - q_{m0} + (1 + \varepsilon) q_{rm} - (1 + \varepsilon) q_{mr}/\varepsilon]}{\left(q_{r0} + \frac{q_{m0}}{\varepsilon} \right)(q_{0r} - \varepsilon q_{0m})}.$$

In this expression the parameter ψ allows for the influence of the emission of spontaneous radiation mainly due to the transitions between the laser levels. If the electron density is so high that we can ignore all the spontaneous transitions, i.e., if we satisfy the additional condition

$$n_e \gg \frac{A_r + \varepsilon A_{rm} - A_m}{q_{r0} - q_{m0} + (1 + \varepsilon) q_{rm} - \frac{1 + \varepsilon}{\varepsilon} q_{mr}}, \qquad (21)$$

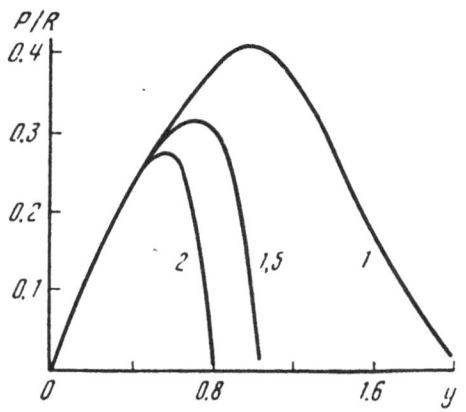

Fig. 4. Time dependences of the saturated power obtained ignoring all spontaneous transitions; the values of the parameter k are given alongside each curve.

the expression for the saturated power becomes

$$P = Ry[1 - k(1 - e^{-y^2})]. \tag{22}$$

The dependence $P/R(y)$ obtained for this case is plotted in Fig. 4. The maximum saturated power P_{max} and the position of this maximum on the time axis are governed by the parameter k. The dependences of these quantities on k are plotted in Fig. 5. The influence of allowance of the term with ψ is illustrated in Fig. 6, which gives the dependences $P/R(y)$ for k = 1 (Fig. 6a) and k = 2 (Fig. 6b), for different values of ψ.

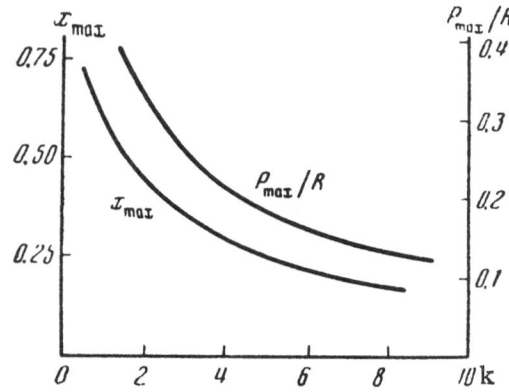

Fig. 5. Dependences of the maximum saturated power and of the time taken to reach this maximum on the parameter k in the absence of spontaneous transitions.

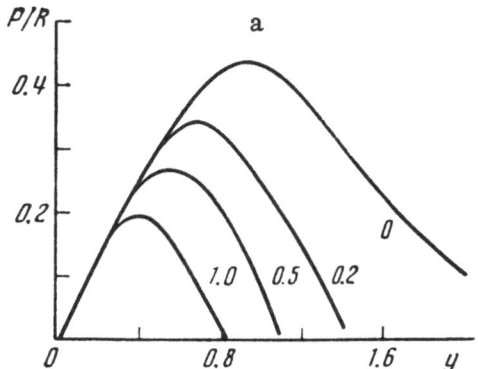

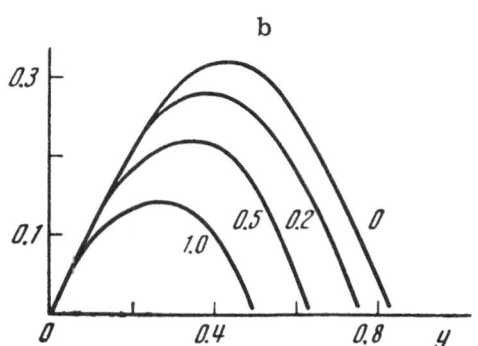

Fig. 6. Influence of the spontaneous emission as a result of transitions between the active (laser) levels. The numbers alongside the curves give the values of ψ.

§ 5. Pulse Superradiance due to Transitions
from Resonance to Metastable Levels

We shall now consider the case of practical importance when pulse superradiance is emitted as a result of transitions from resonance to metastable levels. Such transitions have a number of advantages, mentioned in the Introduction. Moroever, the formulas for this case simplify somewhat so that they become clearer.

Since the lower active level is metastable, we shall assume that

$$A_r \gg A_m, \quad A_{rm} \gg A_m. \tag{23}$$

Moreover, we shall assume that the excitation of a metastable level by electrons is weak, i.e., we shall postulate that

$$q_{0r} \gg q_{0m}, \quad q_{r0} \gg q_{m0}. \tag{24}$$

We have seen earlier that the nature of decay of the active levels is a very important factor. In the resonance case we have to allow for the reabsorption (trapping) of the resonance radiation. We shall describe the probability of radiative decay of the upper level by

$$A_r = A_{r0}F + A_{rm}, \tag{25}$$

where A_{r0} is the probability of a spontaneous transition from the resonance to the ground level, and $F = F(N_0, r)$ is a factor which allows for the reabsorption (trapping) of the radiation and which depends on the population of the ground level and the dimensions of the region occupied by the active gas [20]. For a cylindrical tube, we have

$$F = \frac{1.60}{k_0 r \sqrt{\pi \ln (k_0 r)}}, \tag{26}$$

where k_0 is the absorption coefficient at the center of a Doppler-broadened line and r is the tube radius.

We shall first consider the case of relatively low electron densities. Using Eqs. (23)–(25), we find that the condition (14) becomes

$$n_e = \alpha t \ll \frac{A_{r0}F + A_m/\varepsilon}{q_{r0}}. \tag{27}$$

This condition may be satisfied or otherwise depending on the factor F. If the reabsorption of the resonance radiation is strong, this condition is violated even for low values of n_e, because in our case A_m is small. We shall consider the case when F is still not too small and the condition (17) is satisfied. Then, $A_{r0}F \gg A_m/\varepsilon$ and

$$P = M [x - (\theta + \varkappa x)(x - 1 + e^{-x})], \quad x = \rho t, \tag{28}$$

where the expressions for the parameters in the above equation depend explicitly on F:

$$\rho = \frac{\varepsilon}{1 + \varepsilon} A_{r0}F; \qquad M = \frac{N_0 \alpha q_{0r}}{\varepsilon F A_{r0}};$$

$$\theta = 1 + \frac{(1 + \varepsilon) A_{rm}}{A_{r0}F}; \qquad \varkappa = \alpha \frac{(1 + \varepsilon)^2}{\varepsilon} \frac{q_{rm} - q_{mr}/\varepsilon}{(A_{r0}F)^2}.$$

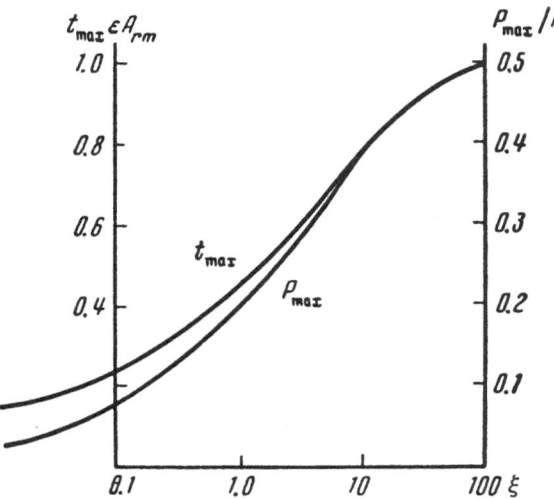

Fig. 7. Influence of retrapping of radiation: the curves represent the dependences of the maximum saturated power and of the time taken to reach this maximum on the parameter ξ.

It is convenient to introduce a dimensionless parameter $\xi=\frac{1+\varepsilon}{F}\frac{A_{rm}}{A_{r0}}$, which depends on the degree of reabsorption of the resonance radiation. We can readily show that in the $\varkappa = 0$ case, we have

$$t_{max} = \frac{1}{\varepsilon A_{rm}} \xi \ln \frac{\xi+1}{\xi},$$

$$P_{max} = \frac{\alpha N_0 q_{0r}}{\varepsilon(1+\varepsilon)A_{rm}} \xi \left[1 - \xi \ln \frac{\xi+1}{\xi}\right]. \tag{29}$$

The dependences of t_{max} and P_{max} on the parameter ξ are plotted in Fig. 7. When the reabsorption increases in importance, both these quantities increase and at $\xi = 10$, i.e., when $(1+\varepsilon)A_{rm} \approx 10A_{r0}F$, the rise practically stops. Thus, if the superradiance power is to be as high as possible, the effective probability of decay of the upper level to the ground state should be considerably less than the probability of the laser transition between the active levels.

This condition is essentially identical with Eq. (13). In cases of practical interest this condition is readily satisfied because the reabsorption of the resonance radiation is relatively weak in active media of the usual dimensions. Thus, calculations can always be made using the simplified formulas of the system (10') in the case of transitions from resonance to metastable levels in media with low electron densities. At high electron densities we have to use Eq. (20) or, if the condition (13') is satisfied, we can employ the simpler formula (12').

§ 6. Saturated Superradiance Power of

Green Thallium Line

We have derived general formulas for the saturated power and now we shall apply them to a specific case. We shall consider the thallium atom because it is typical of a system in which stimulated radiation is obtained as a result of transitions from the first resonance to metastable level.

The energy level scheme of the thallium atom is given in Fig. 8. Stimulated radiation is observed as the 5350 Å line due to the $7^2S_{1/2}-6^2P_{3/2}$ transition (see Chap. III). The probabilities and oscillator strengths of the resonance and active (laser) transitions are $A_{rm} = 6.6 \cdot 10^7$ sec^{-1}, $f = 0.14$, $A_{r0} = 6.8 \cdot 10^7$ sec^{-1}, $f = 0.13$ [21]. At working electron densities

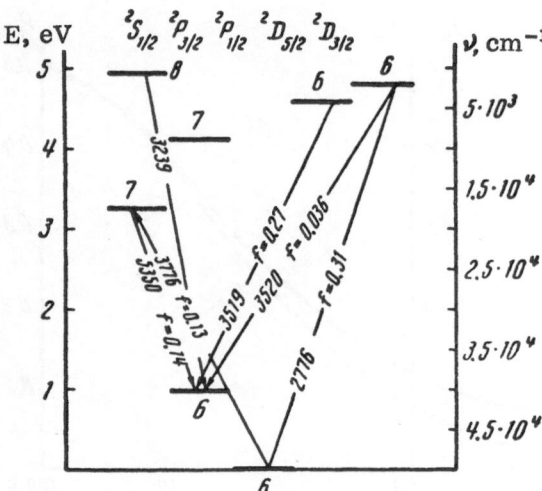

Fig. 8. Energy levels and transitions in the thallium atom.

the resonance-transition reabsorption is so strong that we can ignore the spontaneous decay as a result of this transition. Thus, Eq. (13) is satisfied. Unfortunately, there is no published information on the electron-excitation cross sections of thallium levels. Therefore, we shall assume that the excitation from the ground state to the metastable level and the decay back to the ground state are negligible processes and we shall assume that $q_{0m} \approx q_{m0} \approx 0$. The other values of q_{ik} will be estimated using approximate formulas which are valid for allowed transitions [22]. In particular, we shall assume that

$$\sigma_{max} = \pi a_0^2 \frac{2f}{\varepsilon^2}, \tag{30}$$

where πa_0^2 is the atomic unit of cross section, f is the oscillator strength of a transition, and ε is the energy of a transition expressed in rydbergs. The electron velocity distribution is assumed to be Maxwellian. The values of q_{ik} (in units of 10^8 cm^3/sec) calculated in this way are listed in Table 1 for several electron temperatures. It is clear from Table 1 that the condition (13) is satisfied for low values of T_e and it begins to be violated from $T_e = 4$ eV. Consequently, calculations can be made using simplified formulas.

The populations of the ground and metastable levels of the thallium atom were assumed to be governed by temperature. Information on the thallium vapor pressure was taken from [23]. It was assumed that the initial population of the metastable level was the equilibrium value. For example, at 800°C the population of the ground state was $N_0 = 8.7 \cdot 10^{14}$ cm^{-3} and that of the metastable level was $N_m = 6.8 \cdot 10^{11}$ cm^{-3}. The equilibrium population of the resonance level at the working temperatures was negligible. In the derivation of the general formulas in the preceding sections we assumed that $N_r = N_m$ and $N_m = 0$ at $t = 0$. However, under real conditions one should allow for the initial population of the metastable level of

TABLE 1

kT_e, eV	q_{02}	q_{12}	q_{20}	q_{21}
0.5	0.0042	0.06	3.1	16
1.0	0.13	0.68	3.52	14.6
2.0	0.75	2.28	3.86	14.4
3.0	1.3	3.26	3.9	14.1
4.0	1.69	3.8	3.87	13.6

TABLE 2

α, $cm^{-3} \cdot sec^{-1}$	t_{max}, nsec	Results of calculations	Temperature, °C										$n_e(t_{max})$, cm^{-3}
			900	940	960	1000	1040	1060	1100	1140	1160	1200	
		kT_e=1 eV											
10^{20}	30	$N_r \cdot 10^{-10}$, cm^{-3}	0,35	0,70	1,19	2.84							
		$N_m^{equil} \cdot 10^{-10}$, cm^{-3}	0.14	0.51	1.01	3.98							$3 \cdot 10^{12}$
		P, W	0,08	0,17					Absorption				
10^{21}	27	$N_r \cdot 10^{-11}$, cm^{-3}	0,28	0.57	0.96	2.29	4,88	6.67	13.8	29.6	38,7		
		$N_m^{equil} \cdot 10^{-11}$, cm^{-3}	0.014	0.051	0.1	0.4	1.3	2.1	6.8	19.7	31,7		$2.7 \cdot 10^{13}$
		P, W	0.79	1,58	2.65	6.35	13,5						
10^{22}	18	$N_r \cdot 10^{-12}$, cm^{-3}	0,13	0.26	0.43	1.03	2,19	3.0	6,21	13,3	17.3	25.9	
		$N_m^{equil} \cdot 10^{-12}$, cm^{-3}	0.001	0.005	0.01	0.04	0,13	0.21	0,68	1.9	3.5	6.3	$1.8 \cdot 10^{14}$
		P, W	5.8	11.8	20	47	100	137	285	610	765		
10^{23}	9	P, kW	0.03	0,05	0.09	0.21	0.46	0,63	1.94	3.65	4.4	5.45	$9 \cdot 10^{14}$
		kT_e=3 eV											
10^{20}	30	$N_r \cdot 10^{-11}$, cm^{-3}	0,35	0.70	1,18	2.82	6.0	8.5					
		$N_m^{equil} \cdot 10^{-11}$, cm^{-3}	0.01	0.05	0.1	0.4	1.0	2.1					$3 \cdot 10^{12}$
		P, W	0.84	1.7	2.8	6,8	10.3	20		Absorption			
10^{21}	28	$N_r \cdot 10^{-12}$, cm^{-3}	0.3	0.6	1.02	2.4	5.12	7.0	21.6	40.6	49,1	204	$2.8 \cdot 10^{13}$
		$N_m^{equil} \cdot 10^{-12}$, cm^{-3}	0.00	0,00	0.01	0,04	0.1	0.13	0.68	2,0	3,1	6,3	
		P, W	7,9	15.7	26.5	63.5	135	185	570	1070	1290	1590	
10^{22}	20	P, kW	0.06	0,12	0.21	0.50	1.07	1.47	3.04	6.5	8.5	12.6	$2 \cdot 10^{14}$
10^{23}	9	P, kW	0.33	0.68	1.13	2.7	5.8	7,9	24,3	46.7	55	68	$9 \cdot 10^{14}$

thallium. In particular, since the initial population of the metastable level of thallium was quite significant, population inversion and stimulated radiation could be achieved only when a certain threshold rate of pumping of the resonance level was exceeded. This was due to the fact that in a time of the order of the lifetime of the upper active level one would have to establish a population of this level exceeding the equilibrium population of the metastable level. Thus, α should exceed a certain definite value. The calculations were carried out as follows. We determined N for a given value of α. The saturated power was calculated only for $N_r \gg N_m^{equil}$. Table 2 gives the results of calculations carried out for different values of T_e and α; it lists also the equilibrium population of the metastable level at a given temperature and the calculated population of the resonance level at the moment when the power maximum was reached. The saturated power was not calculated when the populations were of the same order of magnitude. Moreover, we included in Table 2 the values of t_{max} (nsec) and $n_e(t_{max})$.

It is clear from Table 2 that, under the assumed conditions, stimulated radiation should be obtained only for such values of α and n_e for which transitions between the active levels may occur as a result of collisions with electrons. Consequently, the time needed to reach the maximum power depends on the electron density and it is considerably shorter than the value calculated allowing for the spontaneous decay (in the latter case t_{max} = 30 nsec). The saturated power depends strongly on the electron temperature and on the rate of rise of the electron density. For $\alpha = 10^{23}$ cm$^{-3} \cdot$ sec^{-1} saturated power may reach tens of kilowatts per cubic centimeter. Thus, calculations show that thallium can be used as an active medium in a pulse laser which should emit a considerable power in the visible part of the spectrum. As pointed out earlier, the laser should be characterized by a high efficiency.

Thus, a calculation carried out for a typical (thallium) laser utilizing transitions from resonance to metastable levels shows that in order to achieve a significant peak output power we have to ensure that the values of α and n_e are such that the decay (deexcitation) of the upper level is governed by collisions of electrons and occurs in the laser channel (as a result of the transition between the active levels). The duration of stimulated radiation pulses decreases with increasing α but the peak power rises. This situation should be typical of lasers utilizing transitions from resonance to metastable levels.

Our calculation also shows that population inversion and stimulated emission last only for a limited time and they are followed by strong absorption. The duration of stimulated emission pulses depends on many parameters and it decreases for large values of n_e, which should be allowed for in analyses of experimental results.

CHAPTER II

APPARATUS AND MEASUREMENT TECHNIQUES

§ 1. Gas-Discharge Tubes and System for
Evacuation and Filling with Gases

In our experiments the active medium was a pulse-discharge plama in various gases, which was generated in fused quartz (vitreous silica) or molybdenum glass tubes of different lengths and diameters. When high discharge current densities were needed in a small volume, we used capillary tubes 20 cm long and with an internal diameter of 1.3 mm. In some cases we employed wider discharge tubes of up to 10 mm in diameter and up to 120 cm long. All the discharge tubes had optical-quality quartz windows oriented at the Brewster angle.

Longitudinal discharges used in our experiments were excited with two cold electrodes sealed into side tubes. The stimulated output often increased when the cathode was a hollow cylinder. Therefore, in all our discharge tubes the cathode was a hollow cylinder. The anode in quartz tubes was a massive tungsten electrode taken from an IFP-2000 lamp; in glass tubes we used a Kovar cylinder. No special studies were made of the influence of the electrode shape and material on the stimulated emission. A typical discharge tube is shown in Fig. 9a.

The discharge conditions could be varied readily because the tubes were connected to a glass vacuum system which had a TsVL-100 oil diffusion pump or an SDN glass pump.

Before measurements a discharge tube was outgassed and conditioned by a neon discharge. This conditioning, followed by second evacuation, ensured stable results in studies of laser emission. The degree of outgassing of the discharge tubes could be judged roughly by the color of a discharge neon. A more accurate check was obtained by a visual examination of the neon discharge spectrum. Molecular bands were found in this spectrum if a tube was poorly outgassed. An additional check of satisfactory outgassing was the observation of some known laser emission or superradiance lines. It was noted that the intensity of the 6143 Å superradiance line of neon was very sensitive to the purity of the gas and this line was observed in a well-outgassed tube. Therefore, the degree of outgassing of narrow tubes was judged by the 6143 Å neon superradiance line.

We used inert gases of spectroscopic purity and high-purity metals. Before the admission of a gas, a tube was evacuated using liquid-nitrogen traps until 10^{-6}-$5 \cdot 10^{-6}$ Torr was reached. The pressure of the gases introduced into a discharge tube was measured with an oil manometer. In those cases when the active medium was a metal or salt vapor, the necessary gas pressure was reached by heating and it was monitored by measuring the temperature. We used tubular heaters made in our laboratory as follows. An asbestos-cement tube 30-50 cm long was wound with a Nichrome wire 1 mm thick. The external heat insulation was provided by an asbestos sheet. The ends of the heater were closed by asbestos-cement covers which had apertures for the insertion of the discharge tube. A temperature of 1000°C was easily reached in such a heater and the power consumption of the heater was 2 kW. The energy losses due to heating could be lowered by reducing the heater diameter. However, heaters of small diameter were inconvenient because of the danger of breakdown and capacitative leakage from the discharge tube to the heater.

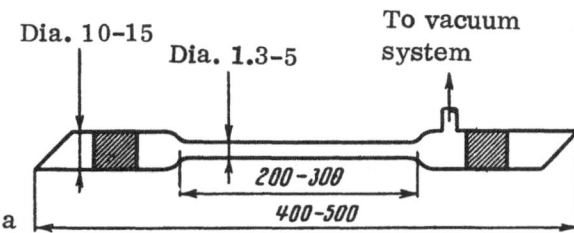

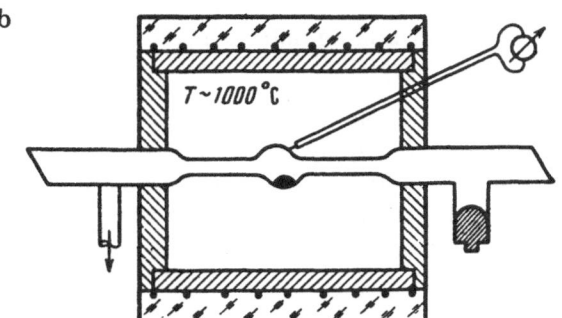

Fig. 9. Discharge tubes: a) used in our investigation; b) surrounded by heater. All dimensions are in millimeters.

The maximum temperature which could be reached in our heaters was 1100°C but this reduced strongly the service life of the heater so that the experiments were carried out at temperatures up to 1000°C. Usually a heater was placed around a discharge tube in such a way as to heat only the central part of the tube keeping the windows and electrodes outside the heated zone. The loss of the active substance from the heated zone was minimized and condensation on the windows was avoided by the use of buffer gases, which ensured that the electrodes were separated from the working part. The buffers were inert gases at pressures of several torr.

An investigated metal was placed in a wider part or in a side tube joined to the central part of the discharge tube. Heating caused diffusion of the metal atoms along the tube and their deposition on the tube walls near the points where it emerged from the heater. Thus, a diffusion flux and a gradient of the active atom density existed along the tube. The density of the active atoms in the central part of the tube was governed by the temperature of its walls. In all experiments the temperature was measured outside the tube near the wider part or the side tube; we used a thermocouple or a thermometer. A typical discharge tube inside a heater is shown in Fig. 9b.

§ 2. Resonators, Mirrors, Superradiance

Almost all the systems which we investigated had such a high gain that superradiance was observed readily. The radiation emitted from long discharge tubes resembled that obtained from a conventional laser with a resonator. Superradiance was not only directional (for a given line) but also exhibited an increase in the intensity of the line which stood out from the spectrum. The emission of superradiance could be deduced also from the time characteristics of the emission of a given line because in all cases investigated by us the duration of stimulated emission was much shorter than the emission of spontaneous radiation. In the visible part of the spectrum the appearance of superradiance could easily be detected visually even near the threshold because of the specific structure exhibited by stimulated radiation.

Additional criteria of the appearance of superradiance were the line narrowing and the nature of the dependence of the line intensity and duration of emission on the experimental conditions (discharge current, gas pressure, applied voltage, etc.). All these characteristics differed sharply from those of spontaneous radiation: for example, under certain conditions the intensity rose very rapidly and the duration of the output pulses decreased strongly, which indicated the appearance of superradiance. These properties made it possible to detect superradiance without any difficulty.

In some cases superradiance was observed clearly only in the presence of one mirror, which reflected the radiation back along the discharge tube axis, i.e., ensured that the radiation passed through the discharge tube twice. As explained earlier, we defined superradiance as the stimulated radiation which was observed in the presence of one mirror or without any mirrors. It should be pointed out that use of one mirror not only increased strongly the radiation intensity but also reduced considerably the angular divergence of superradiance. We defined laser action (emission) as that case of stimulated emission when an optical resonator with two mirrors had to be used.

When the laser emission and superradiance were considered from the point of view of population inversion mechanisms, they differed only in respect of the value of the gain. However, it should be noted that in the presence of a resonator the laser radiation acquired a mode structure, governed by the type of resonator and its dimensions. Superradiance was not influenced by the properties of the external mirrors but was governed by the parameters of the active medium itself.

Superradiance was observed easily in our systems. Consequently, we studied the radiation emitted by discharge tubes without resonators. Usually one mirror (with a dielectric coating or aluminized) was placed behind the tube and it was aligned in such a way so that it reflected the radiation back along the discharge tube axis. This alignment was usually made employing the 6143 Å superradiance line, which — under certain conditions — readily formed a strong directional beam after passing twice through the active medium. We used plane mirrors. The degree of influence of the mirror on the emitted radiation was checked by covering the beam with a black screen on the mirror side. The use of a high-reflectivity second mirror usually only reduced the intensity of the output beam. In many cases it was not possible to achieve optimal reflection by the second mirror because of the absence of a sufficiently wide range of mirrors with different reflection coefficients.

The measured inversion lifetime was found to be of the order of several nanoseconds, i.e., it was comparable with the time taken by light to travel between the mirrors. In this case a resonator would have been practically useless and one could only obtain superradiance. Clearly, even in the presence of one mirror, its influence could depend on the distance between the mirror and the active medium. Such an influence was indeed observed experimentally. This effect and the others mentioned above will be discussed in detail later.

§ 3. Pulse Power Supply System

In the course of our investigation it was found that many of the investigated lasers emitted at the beginning of the current pulse and the duration of stimulated emission was only a few nanoseconds or less. In such cases the excitation pulses should be short and should have a steep leading edge.

A pulse cable transformer system [24] was used for the first time in our study as a power supply source of pulse self-terminating lasers. The system was found to be relatively simple to construct and very convenient to use. It is now employed extensively in pulse lasers [25].

This power supply system is shown in Fig. 10 together with the discharge tube. The main part of the system was a pulse cable transformer whose primary winding was excited by discharging a charge line (Fig. 10a) or a capacitor (Fig. 10b) through a thyratron. The secondary winding of the transformer was connected to the discharge tube electrodes.

Our transformer was made of ten ferrite rings of the F-1000 type: their internal diameter was 80 mm and the external diameter was 120 mm. Twelve turns of an RK-106 cable were wound on the ferrite-ring core. The cable braid, which was used as the primary winding, was divided into three sections of 3.5 turns each and these sections were connected in parallel. The secondary winding was the core and it consisted of 12 turns. Thus, the turn (transformer) ratio was 3.5. The maximum voltage on the primary winding was governed by the type of thyratron employed and by the electric strength of the cable. In our case the voltage applied to the primary winding could be varied from 0 to 16 kV and the stepped-up voltage produced by the secondary winding could reach 50 kV.

A two-stage symmetric line (Fig. 10a) was made of an RK-49 coaxial cable section 5 m long, which corresponded to a 50-nsec pulse duration in a matched load. The internal resistance of the power supply system with this line was given by the formula

$$R_i = 2\rho k^2,$$

where ρ is the wave resistance of the line and k is the turn ratio. Parallel connection of n coaxial cables with a wave resistance ρ_0 made it possible to reduce the internal resistance of

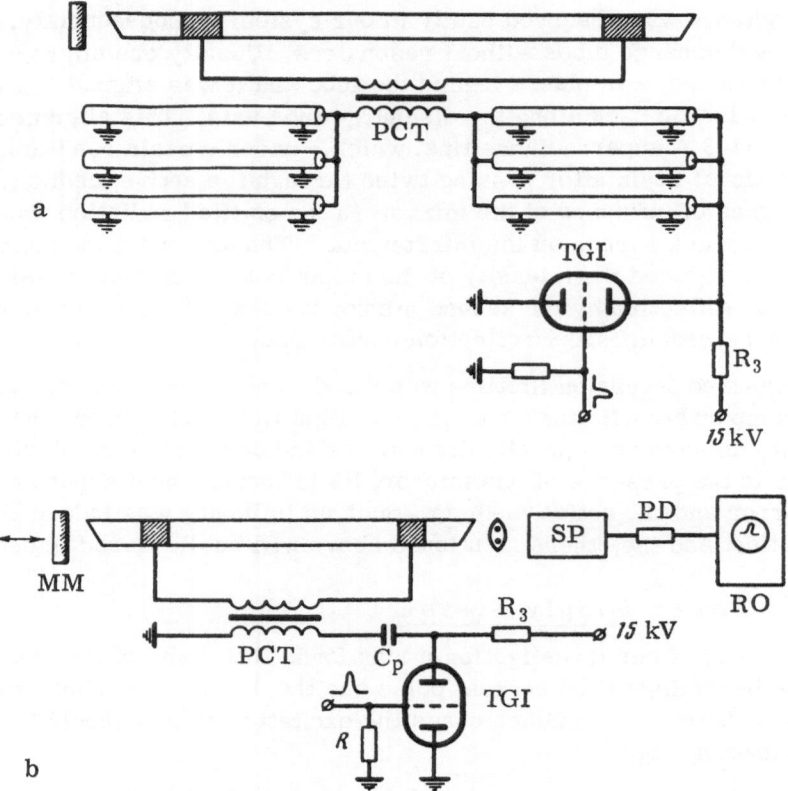

Fig. 10. Power supply system with a pulse cable transformer (PCT): a) circuit with a symmetric charging line; b) circuit with a capacitor C_p as energy storage device. Here, MM is a movable mirror: SP is a spectroscopic instrument; PD is a photodetector; RO is a recording oscillograph; TGI is a switching thyratron.

the system in accordance with the formula $\rho = \rho_0/n$. We used five cables with $\rho_0 = 75\ \Omega$ connected in parallel so that the internal resistance was 367.6 Ω.

We also used a different excitation system (Fig. 10b) for obtaining laser action and superradiance as a result of self-terminating transitions. In this case we discharged a capacitor through the primary winding. We used mainly KVI-3 low-inductance capacitors of 0.001, 0.004, and 0.01 μF capacitance. The parameters of the laser emission and superradiance were not greatly affected by the actual capacitance employed or by the power supply variant used. The power supply system was exceptionally simple and compact; moreover, it ensured a high output and was reliable.

Different thyratrons (for example, TGI1-400/16, and TGI1-130/10 or, during the last stages of this investigation, TGI1-500/16) were used as the switches. The pulse repetition frequency could be varied from a few hertz to several kilohertz. It was limited by the capabilities of the power rectifier and the nature of the thyratron employed. The TGI1-130/10 thyratron was characterized by a steeper leading edge [26], because the results obtained using this device were not inferior to those obtained employing other thyratrons operating at higher voltages.

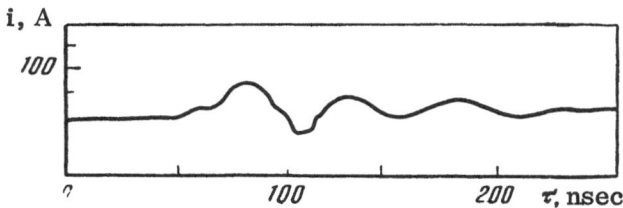

Fig. 11. Typical oscillogram of a current pulse in a discharge tube.

A very simple circuit generated pulse pairs. A switching thyratron was triggered by pulse pairs produced by a GIS-2 generator. The delay between these pulses could be varied from 300 to 2000 μsec.

The power supply system described above produced current pulses with a leading edge of 15-20 nsec and a total duration of 100-150 nsec. These pulses were in the form of a damped oscillation with a maximum amplitude of 200-300 A. A typical oscillogram of a current pulse in the discharge tube is shown in Fig. 11.

§ 4. Measurement of Spectral and Time Characteristics

Spectral and time characteristics of the spontaneous and stimulated radiation were studied particularly thoroughly. The physical processes in an active gas-discharge plasma were investigated mainly by recording and analyzing the stimulated and spontaneous emission spectra and their changes with the experimental conditions. The time characteristics, i.e., the position on the time axis, shape, and duration of the stimulated radiation pulses representing individual lines, were determined in order to identify the population inversion mechanisms.

We employed the following spectroscopic apparatus for recording the spectra and separating individual lines. The visible and ultraviolet spectra were studied with an STÉ-1 echelette spectrograph, which enabled us to photograph spectra with a high dispersion in a wide spectral range. This spectrograph was used mainly in the search for new laser emission lines and for photographing the spontaneous radiation spectra. The wavelength could be determined to within 0.06 Å. When a new stimulated emission wavelength has to be measured accurately, we used a DFS-13 spectrograph with a focal length of 4 m and with a first-order linear dispersion 4 Å/mm for a grating with 600 lines/mm and 2 Å/mm for a grating with 1200 lines/mm. In this case we were able to determine the wavelengths to within 0.03 Å. This precision was sufficient to identify reliably the investigated lines.

In addition to photographic recording, we used photoelectric recording in studies of the spectra and particularly in investigations of the shape and duration of the radiation pulses representing individual lines. We used mainly a DFS-12 double monochromator with two diffraction gratings (600 lines/mm) and a linear dispersion 5 Å/mm in the working order; moreover, in the photoelectric variant we employed ISP-51 and IKS-6 spectrographs.

Photoelectric recording was used exclusively in the infrared range.

Hyperfine structure of the stimulated emission lines and the widths of the superradiance and laser emission lines (as well as other properties) were studied with IT 51-30 or IT 28-150 Fabry−Perot interferometers.

A considerable stress was laid on the investigation of the time characteristics. Since the duration of the investigated pulses was very short (of the order of nanoseconds for stimulated radiation and tens of nanoseconds for spontaneous radiation) and the power was relatively low, the determination of the time characteristics was a difficult experimental task. Initially we used apparatus with a time resolution of the order of 10-20 nsec in the visible and ultraviolet parts of the spectrum. It was frequently found that the measured duration of superradiant pulses was of the order of the time resolution of the recording apparatus. Only during the later stages of the investigation were we able to use detectors with a sufficient sensitivity and a time resolution of 1-0.5 nsec.

In the majority of the measurements carried out in the visible part of the spectrum the output of a spectroscopic instrument was passed to an FÉU-36 photomultiplier whereas an FÉU-28 photomultiplier was used in the infrared range. Linear operation of the photomultipliers was ensured by weakening, if necessary, the light flux with gray filters. The photomultiplier signal was applied to an amplifier, which was connected to an S1-11 oscillograph. The duration of the superradiance pulses found in this way was about 10 nsec. This was equal to the time resolution of the recording apparatus and it was governed by the pass band of the oscillograph amplifier and by the photomultiplier itself. The sensitivity of the recording system, consisting of the FÉU-36 or FÉU-28 photomultiplier and the S1-11 oscillograph, was sufficient for recording of the majority of the spontaneous radiation lines in the visible, ultraviolet, and near infrared parts of the spectrum. Moreover, the time resolution was sufficient for the measurements of the spontaneous radiation.

Stimulated radiation pulses were recorded with a time resolution of about 1 nsec using an FÉK-16 coaxial photocell. The signal from this photocell was applied to the plates of an I2-7 time interval meter. The general arrangement used was of the type shown in Fig. 12. This apparatus was capable of recording superradiance pulses of 0.5-0.7 nsec duration at midamplitude. Thus, the photomultiplier measurements gave the energy of the stimulated radiation pulses, whereas the coaxial photocell measurements gave the superradiance and laser emission power. The coaxial photocell could be used to record green light pulses with a peak power of 10 W. The sensitivity of this cell was insufficient for superradiance measurements near the threshold. Therefore, we used special high-current time multipliers of the SNF or SNFT type. Their sensitivity was between four and eight orders of magnitude higher than that of the FÉK-16 autocell and their resolution was about 3 nsec.

The apparatus described above provided satisfactory means for investigating the time characteristics of the stimulated radiation pulses in the visible and ultraviolet parts of the spectrum. The time resolution in the infrared range was poorer and it was insufficient for the determination of the duration of superradiance pulses at some wavelengths.

Very strong electrical stray signals, which could interfere considerably with the investigated processes, were generated as a result of high-power pulse discharges for the excitation of gas lasers when the excitation source was not matched exactly to the discharge tube. The influence of these strays was minimized by double screening of all the coaxial cables (carrying signals representing various processes) and by screening of the photomultiplier power supply. The photomultiplier itself was carefully enclosed in a thick metal tube. Special attention was paid to the quality of all the contacts in the screen and signal circuits. Special measures were taken to prevent penetration of strays via the line supply (these measures included the use of narrow-band filters). The whole system was grounded at the same point and this point was selected by trial and error so as to minimize the stray signals. When all these precautions were taken, it was possible to obtain undistorted signals on the oscillograph screen. It should

Fig. 12. Schematic diagram showing a coaxial photocell (FÉK-16) used to measure the shape and duration of superradiance pulses generated in a discharge tube (DT).

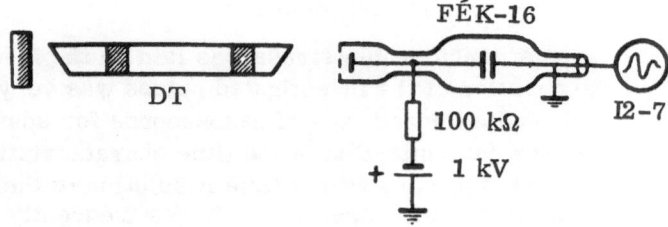

be mentioned that the strays were much less important when the signal representing the investigated process reached directly the oscillograph plates than then when it passed first through the amplifier. This was particularly true of the arrangement with the coaxial photocell when the strays had least influence anyway.

The stimulated radiation power was measured by two methods. A calibrated thermopile was used to determine the average power at selected pulse repetition frequency. Next, the known pulse duration and repetition frequency could be used to calculate the peak power. In the second method we carried out absolute calibration of a coaxial photocell and deduced the peak power directly from the signal produced by this photocell. For example, at the 5350 Å wavelength the photocell had a sensitivity of 21.6 mA/W. The results obtained by these two methods were in satisfactory agreement.

§ 5. Measurement of Electrical Characteristics of Discharges

The knowledge of the duration and shape of the current pulses, their amplitude, and of the shape of the applied voltage pulses was necessary for the understanding of the physical processes resulting in population inversion and for the determination of the efficiency of a given type of laser, and in many other tasks. It was quite difficult to measure current pulses with such very short leading edges (amounting to tens of nanoseconds).

Measurements on current pulses were carried out using a coaxial shunt and a Rogowski loop operated as a current transformer [27]. The coaxial shunt was an MON 1E1I resistance soldered directly into an RK-75-4-21 cable, as shown in Fig. 13. The shunt was checked and calibrated with the pulse generator of the I2-7 oscillograph which produced pulses of ~10-100 nsec duration. A current pulse produced by this generator and recorded with our shunt is shown in Fig. 14, where a is a pulse leaving the generator and b is the same pulse received using the shunt whose signal was applied to the amplifier of the S1-11 oscillograph. The shunt

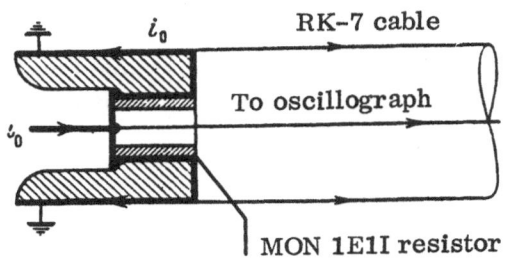

Fig. 13. Coaxial shunt based on an MON 1E1I resistor.

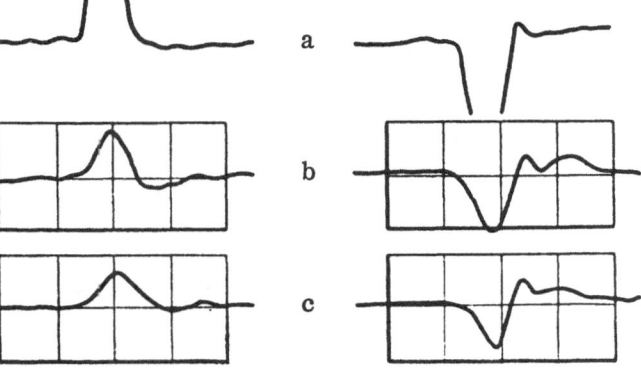

Fig. 14. Oscillograms of ~10 nsec calibration pulses (those of the negative polarity are on the right and those of the positive polarity on the left): a) pulse at the generator output; b) pulse recorded with a shunt; c) pulse recorded with a Rogowski loop. Each division represents 20 nsec.

resistance was measured using ~10 nsec pulses and it was 1.17 Ω; it was practically the same as the low-frequency value. Hence, we concluded that the skin effect had no influence on the shunt measurements. Consequently, the shunt was linear within the range 0-100 MHz.

Current pulses in the ungrounded part of the circuit were studied using a Rogowski loop operated as a current transformer. One turn, which was short-circuited by a coaxial resistance (described above), was wound on a ferrite ring. The inductive impedance of the measuring circuit was then higher than the resistance. Hence, the signal produced by the loop was [27]

$$u = i \frac{R}{n},$$

where R is the loop resistance, n is the number of turns, and i is the measured current. In our case we had n = 1 and, therefore, u = Ri. The Rogowski loop operation was checked and calibrated in the same way as that of the shunt. Figure 14c shows oscillograms of calibration pulses obtained using the Rogowski loop.

The shape of the voltage pulses applied to the discharge tube was determined using an ohmic voltage divider composed of inductance-free TVO resistors. This divider was checked and calibrated in the same way as the current source.

The current and light pulses were observed simultaneously by applying two signals, one from the current source and the other from the photodetector, to the I2-7 oscillograph plates via a matched mixer. Conditions were selected so that the time taken by the signals to pass along the channels was the same to within 1 nsec.

No reliable methods were available for the determination of the electron density and temperature in discharges excited by the very short pulses used in our study. The situation was additionally complicated by the fact that the plasma parameters at the moment of stimulated emission, i.e., in a time of the order of 1 nsec from the beginning of the excitation pulse, were needed in the interpretation of population inversion mechanisms. Such parameters were difficult to measure and we did not attempt to determine them.

The relatively simple apparatus described above enabled us to observe and investigate many new superradiance and laser emission lines due to self-terminating transitions, as well as all the lines reported in the literature.

CHAPTER III

PULSE LASER EMISSION AND SUPERRADIANCE DUE TO TRANSITIONS FROM RESONANCE TO METASTABLE LEVELS IN ATOMS AND IONS

§ 1. Superradiance due to Transitions in Helium Atoms

The helium atom is the simplest system in which transitions from resonance to metastable levels can occur. The atomic properties needed in the analysis of the population inversion processes have been studied more thoroughly for helium than for other atoms or ions of interest to us. In particular, the probabilities of a large number of transitions and the lifetimes of many levels are well known for helium [28]. A large amount of experimental and theoretical work has been done on the electron-excitation cross sections. Finally, experiments on helium are very easy to carry out. In view of these circumstances, it was natural to begin investigation of population inversion processes from transitions between resonance and metastable levels in helium.

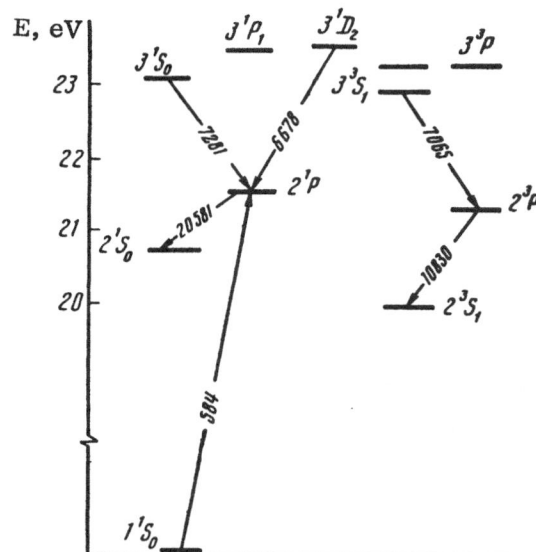

Fig. 15. Simplified energy level scheme
of the helium atom.

The energy level and transition scheme of the helium atom is relatively simple. Its simplified form is shown in Fig. 15. There are practically no intercombination transitions in helium. Continuous laser emission due to two transitions between high-lying helium levels 4^3P-3^3D ($\lambda = 19,543$ Å) and 7^3D-4^3P ($\lambda = 20,603$ Å) were reported [29] by the time we started out study. Population inversion processes between these transitions were not directly relevant to the mechanism which we investigated.

A general analysis of population inversion schemes suggest that efficient electron excitation could be expected particularly for the 2^1P_1 level and that pulse population inversion should occur for the $2^1P_1-2^1S_0$ ($\lambda = 20,581$ Å) transition. The excitation cross section of the upper level of this transition, estimated using the simplified formula (30), is $2.5 \cdot 10^{-17}$ cm^2. The ultimate efficiency of this transition is relatively low, $\eta_{ult} = 0.7\%$, because the 2^1S level is fairly high. The probability of the 2^1P-2^1S transition is $2 \cdot 10^6$ sec^{-1}.

The possibility of obtaining population inversion for this and other transitions in helium is governed primarily by the value of the cross section for the excitation of a given level by electron impact from the ground state of the atom. In the initial stage of our investigation we started from the results given in [30]. A calculation of the populations of the atomic levels of helium based on these data [31] indicated that we could expect pulse population inversion at the beginning of excitation pulses for two transitions: 2^1P-2^1S ($\lambda = 20,581$ Å) and 2^3P-2^3S ($\lambda = 10,830$ Å). Consequently, we carried out experiments designed to detect stimulated radiation as a result of these transitions. In these experiments we observed superradiance for the first of these transitions but neither laser emission nor superradiance were observed for the triplet transition in spite of considerable variation of the experimental conditions [31].

New information is now available on the electron-excitation cross sections of helium levels so that the possibility of population inversion can be reconsidered. The data on the excitation cross sections of the levels of interest to us are given in Table 3. This table gives the excitation thresholds of the levels and the cross sections (in units of 10^{-19} cm^2) obtained by different workers and these quantities are listed for various electron energies. Considerable difficulties are encountered in measurements of the excitation cross sections of various transitions. The transitions from the 2P levels correspond to infrared or far ultraviolet parts of the spectrum and this makes it difficult to measure the cross sections by the usual optical method. The method is altogether unsuitable for the measurement of the cross sections of metastable

TABLE 3

E,eV	2^1P_1 (21,21 eV)							2^1S_0 (20,61 eV)							
	[30]	[32]	[33*]	[34*]	[35*]	[39]	[42]	[30]	[32]	[33*]	[34*]	[38m]	[40]	[41]	[42]
21												5	20		23
23												10	30		
25		17						30					29	10	
27													28		25
30		40	40				64	40	18.3	6,3			25	17	
35		50	44				86	45	24.4	17.8					
40	60	70	49				96,5	43	21.3	23,4				22	
45		85	60							22					
50		110	69	97.5			114	40	18.3						
60	110	125	83	108	70.9		123	38						23	
80	140	150	93	98.7	89.5		123	35		13.7	16			24	
100	150	156	100	87	91.8	120	114	30	12.2	12	12,5			25	
200		125	90	71	79.8		87			7.5	7,6				

F,eV	2^3P (20,96 eV)					2^3S (19.62 eV)								
	[30]	[32]	[34*]	[39]	[42]	[30]	[32]	[33*]	[36m]	[37m]	[38m]	[40]	[41]	[42]
20.5						20			40	26	30			
21						20			30		25	35		
23					55							40	36	60
25	180					61		6.6				50	39	53
30	220	27			44	42	40	30				40	36	42
35		30		80	31		32	40						
40	195	24			22	28	27						26	32
45		18					20							
50	165	15			12	18							10	19
60	135	10			7			15.5					15	12
80		6	4.6		3.1	12	5	4					13	6
100	96	2,6	3.5		1,6		3.5	3					12	3

levels. In view of this excitation cross sections for some levels were found in [30, 32] by extrapolation of the data available for higher levels. A similar extrapolation was applied by us to the results given in [33-35]. We assumed that the energy dependence was the same for all the levels, which was confirmed experimentally. The results obtained in this way are identified by asterisks in Table 3. The cross sections of metastable helium levels were measured by detection of the most metastable atoms knocked out by electrons from solid targets [36-38]. The values obtained in this way are denoted by "m" in Table 3. The results quoted in [39-41], also included in Table 3, were obtained by somewhat different methods. The excitation cross sections of helium have been calculated theoretically by a variety of approximate methods. Table 3 gives only the results of the most comprehensive calculations [42].

A detailed discussion of the experimental and theoretical methods for the determination of the electro-excitation cross sections is given in the review [43] but it does not include the latest data. The possible sources of errors in the determination of these cross sections are also given in [43].

It is clear from Table 3 that the absolute cross sections and their energy dependences obtained by different workers disagree quite considerably. In particular, the results given in [30] for the 2^3P level are greatly overestimated. The values reported by other workers indicate that we can hardly expect inversion for the 2^3P-2^3S transition. It is also clear from Table 3 that the excitation cross section of the 2^1P level is considerably greater than that of 2^1S. Therefore, we can expect a pulse population inversion due to the 2^1P-2^1S transition at the beginning of an excitation pulse.

Superradiance emitted from helium was investigated using a pulse cable transformer described generally in Chap. II. We used two fused-quartz tubes of 1.3 and 5 mm diameter with windows oriented at the Brewster angle. The active length of the discharge was 20 cm. A tungsten electrode taken from an IFP-2000 lamp was used as one electrode and a tubular Kovar or heated oxide cathode from a TGI1-90/8 thyratron was used as the other electrode. The heating current of the oxide cathode was 6-7 A under normal conditions. Spectroscopically pure helium was used in the tube. A discharge was started using a pulse cable transformer with a turn ratio of about 3.5. The voltage applied to the primary winding was varied from 0 to 16 kV. The pulse repetition frequency was varied from 2 Hz to 10 kHz.

Radiation emitted in the 2.058 μ wavelength range was investigated using silvered mirrors and that emitted at 1.08 μ using mirrors with multilayer dielectric coatings and a reflection coefficient of about 99% (in the wavelength range under investigation). A resonator with two mirrors was used only in studies of laser emission. Superradiance at $\lambda = 2.058$ μ was investigated using one mirror or no mirrors at all. The spectroscopic measurements were carried out using IKS-6 and DFS-12 spectrographs. Lead sulfide photoresistors of the FSA-1 type, InSb photodiode, and thermopile were used as radiation detectors. The detector signal was applied to the input of an S1−15 oscillograph for visual examination and photographic recording.

Our experiments revealed superradiance at 2.058 μ. This wavelength could not be measured sufficiently accurately with the IKS-6 spectrograph. Therefore, we identified the transitions responsible for this line by measuring the duration of absorption at the superradiance wavelength in an additional tube. The lifetime of the lower active level (approximately 50 μsec) estimated in this way was far too long to be regarded as the lifetime of any excited helium level, apart from the metastable one. Moreover, measurements carried out using the DFS-12 spectrograph with gratings deflected from the normal position yielded the wavelength 20581.3 ± 0.8 Å. Comparison with tables in [44] indicated that, within the limits of the experimental error, there was no other transition in helium apart from $2\,^1P - 2\,^1S$ which would give this wavelength. Thus, we could definitely say that the observed superradiance was due to the $2\,^1P - 2\,^1S$ transition.

Superradiance pulses emitted from a cold-cathode discharge tube were unstable so that averaging over many pulses had to be carried out in quantitative measurements. The signal produced by an oxide-cathode tube was weaker than that emitted from the cold-cathode tube but it was much more stable. However, the characteristics of superradiance emitted from the oxide-cathode tube depended on the cathode heating current.

It should be mentioned that the energy of superradiance pulses was practically independent of the pulse repetition frequency in the range from 2 Hz to 10 kHz. To the authors' knowledge this was the highest repetition frequency achieved for pulse lasers with self-terminating transitions. Further increase in the repetition frequency was restricted by the power supply system. No significant difference was found between the energy of superradiance pulses generated using different capacitances in the power supply system.

Superradiance characteristics are plotted in Fig. 16. Figure 16a shows the dependence of the energy of superradiance pulses, emitted from a cold-cathode tube of 1.3 mm diameter, on the voltage applied to the pulse cable transformer. The helium pressure was 2.7 Torr. Curve 1 was obtained without a mirror behind the tube and curve 2 with one mirror. Figure 16b gives the same dependence for an oxide-cathode tube. In this case the helium pressure was about 1 Torr. Curves 1 and 2 (with and without a mirror) were obtained using a heating current of 6.2 A, whereas curves 3 and 4 corresponded to a heating current of 5 A. It is clear from these curves that the influence of a mirror on the superradiance emitted from the cold-cathode tube was considerably greater than in the oxide-cathode case. The discharge in the

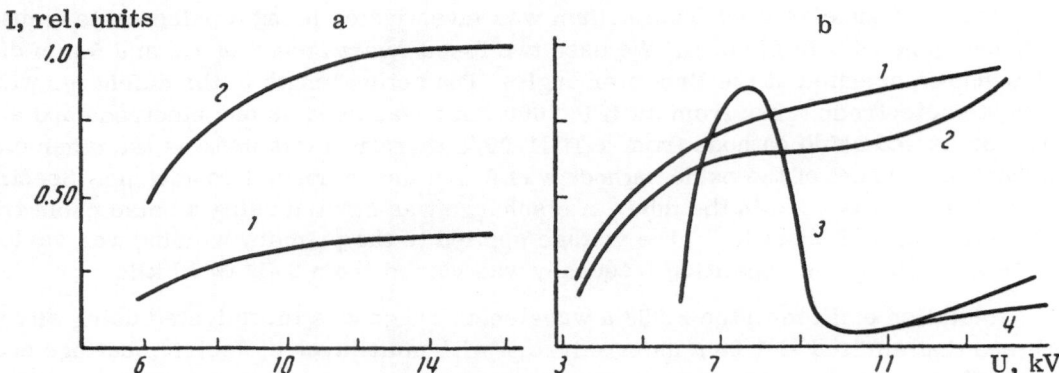

Fig. 16. Dependence of the energy of superradiance pulses emitted from helium on the voltage applied to the cable transformer. a) Cold cathode: 1) without a mirror behind the tube; 2) with a mirror at a distance of 10 cm from the active part of the tube. b) Oxide cathode: 1, 3) with a mirror behind the tube; 2, 4) without a mirror.

oxide-cathode tube began at much lower voltages. The characteristics of the superradiance emitted from the oxide-cathode tube depended very strongly on the cathode heating current. An additional maximum in the dependence of the energy of superradiance pulses on the voltage applied to the transformer was observed when a tube of 1.3 mm diameter was used and the helium pressure was low. In the cold-cathode case this maximum shifted toward higher voltages with decreasing pressure, whereas in the oxide-cathode tube it remained constant for voltages of about 7–8 kV applied to the transformer. The maximum was not observed when a cold-cathode tube of 5 mm diameter was used.

The dependences of the energy of superradiance pulses on the helium pressure are plotted in Fig. 17. The curves in this figure were obtained with one mirror behind the tube. An arbitrary scale was used for each curve. The cathode heating current increased from 6 to 6.7 A when the helium pressure was raised; this was due to the change in the temperature and resistance of the heater when the heat-exchange conditions were altered. When the pressure was lowered, superradiance was emitted from this tube down to pressures at which no discharge took place (of the order of ~0.01 Torr); the discharge and superradiance pulses were unstable.

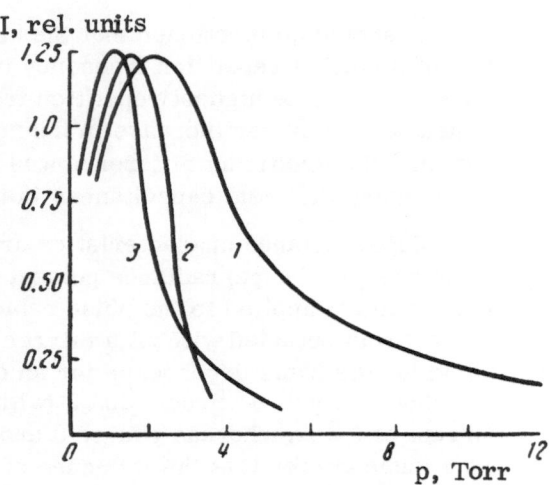

Fig. 17. Pressure dependences of the energy of superradiance pulses emitted from helium: 1) cold-cathode tube of 1.3 mm diameter; 2) tube of 5 mm diameter; 3) oxide-cathode tube of 1.3 mm diameter. The voltage applied to the cable transformer was 12 kV.

We investigated the influence of neon and argon admixtures on the superradiance emitted by helium. The addition of these gases to helium reduced the superradiance power. The helium line superradiance practically disappeared when about 2 Torr neon and ~1.4 Torr argon was added. Neon and argon superradiance lines were emitted from a mixture of helium with neon and argon lines (Chap. IV).

We used an InSb photodiode in a determination of the duration of helium superradiance pulses. Partial cooling of the photodiode with liquid nitrogen (such cooling increased the photodiode sensitivity and also its response time) indicated that the duration of superradiance pulses at midamplitude was 200 nsec. The same detector was used to measure the duration of the 6143 Å neon superradiance pulses, which was 3–5 nsec (Chap. IV). We found that the measured pulse duration (at midamplitude) was again 200 nsec. We concluded that the duration was governed by the response time of the detector and the helium superradiance pulses were considerably shorter.

A calibrated thermopile was employed in measuring the average power of helium superradiance. This power was 0.05 mW for an oxide-cathode tube of 1.3 mm diameter when the helium pressure was 1 Torr, voltage applied to the transformer was 12 kV, and repetition frequency was 5 kHz. This corresponded to an energy of 10^{-8} J per pulse. If the duration of the superradiance pulses was, as usual, of the order of 10^{-8} sec, the peak output power was 1 W. Only the helium pressure was optimal in these experiments; other parameters were not optimized. For example, the energy of superradiance pulses was 10–20 times greater when a cold-cathode tube of 5 mm diameter was used; this energy corresponded to a peak output power of about 20 W. Clearly, when the working conditions were optimized, the transitions in helium could produce a considerable peak power at a high pulse repetition frequency.

We also attempted to observe laser emission or superradiance due to the $2^3P - 2^3S$ transition in the helium atom, for which the expected wavelength was 1.08 μ. We searched for the laser emission using mirrors with multilayer dielectric coatings and a reflection coefficient of 99% in the relevant part of the spectrum. The helium pressure was varied from 10^{-2} to 10 Torr and the voltage applied to the transformer from 0 to 16 kV. No stimulated radiation (neither laser emission nor superradiance) was observed for tubes of 20 cm active length with diameters of 1.3 and 5 mm.

The experimental results indicated that the properties of superradiance due to the $2^1P - 2^1S$ transition were in agreement with those expected on the assumption of direct electron excitation of the active levels from the ground state of the atom. Thus, the general description of population inversion discussed qualitatively earlier was confirmed by the experimental results on helium superradiance and by the available information on the electron-excitation cross sections of helium atoms.

The peak power, pulse duration, and position of pulses on the time axis could not be measured because of the unavailability of a detector with a high time resolution in the wavelength of 2 μ. Consequently, we did not calculate theoretically the superradiance pulse parameters, because experimental data were not available for comparison. Further improvements in the experimental techniques, particularly the development of fast-response detectors sensitive in the relevant part of the spectrum, and refinement of the data on the electron-excitation cross sections of helium levels should result from future investigations. It will then be possible to carry out a reliable comparison of the calculated results with the experimental data.

§ 2. Superradiance of Thallium Vapor

The thallium atom is attracting considerable attention because of its near-ideal level structure and because it satisfies quite well the requirements postulated for high-efficiency

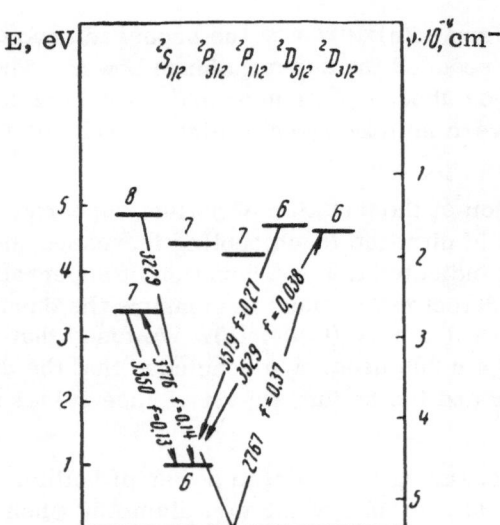

Fig. 18. Energy level and transition scheme
of the thallium atom.

systems. Moreover, it is easier to study experimentally than other atoms with favorable level structures. Moreover, at the time we started our investigation stimulated emission has not yet been obtain from thallium vapor.

The energy level and transition scheme of the thallium atom is given in Fig. 18. Thallium has two resonance levels $7^2S_{1/2}$ and $6^2D_{3/2}$, coupled to the ground state by strong lines at 3776 and 2767 Å. The oscillator strengths of these lines are 0.14 and 0.37, respectively [21]. We were unable to find any published information on the electron-excitation cross sections of thallium levels. An estimate obtained on the basis of the approximate formula (30) gave $\sigma_{max}(7^2S_{1/2}) = 3.8 \cdot 10^{-16}$ cm^2 and $\sigma_{max}(6^2D_{3/2}) = 5.4 \cdot 10^{-16}$ cm^2. It was difficult to estimate the excitation cross section of the metastable level. Bearing in mind the distribution of the electron levels and the fact that the level $6^2D_{3/2}$ had a somewhat higher excitation energy, we could conclude that the rates of excitation of both resonance levels were practically identical.

If we assumed, in accordance with general ideas on population inversion, that the population of the metastable level due to electron impact was negligible, we found that inversion should occur for two transitions terminating at the metastable level: $7^2S_{1/2} - 6^2P_{3/2}$ at $0.535\,\mu$ and $6^2D_{3/2} - 6^2P_{3/2}$ at $0.353\,\mu$. The oscillator strength of the former line was 0.13 and it was considerably greater than the oscillator strength of the latter line, which was 0.038 [21]. Consequently, the gain of the $0.535\,\mu$ line should be considerably higher. Stimulated emission of the $0.353-\mu$ line might not be observed because of the competition resulting from the sharing of the lower level with the $0.535\,\mu$ line. The line starting from the $6^2D_{5/2}$ level with a wavelength $0.352\,\mu$ and a large oscillator strength 0.27 also terminated at the metastable level $6^2P_{3/2}$. However, a direct population of this level by electrons involved a forbidden transition and it was much less effective than the population of the resonance levels. In principle, we could expect the population of the $6^2D_{5/2}$ by collisional transfer of energy from $6^2D_{3/2}$. The difference between the energies of these two levels was small ($\Delta E = 80$ cm^{-1}) but nevertheless the process was not very efficient [45].

Thus, in the first place we should expect stimulated emission of the green line of thallium. The metastable level of thallium is relatively low, being located 7792 cm^{-1} above the ground state. Consequently, the ultimate efficiency of the transition responsible for the green line is very high: $\eta_{ult} = 47\%$. We shall see later than this is the highest value of η_{ult} for all the transitions which have been found experimentally to give rise to laser emission. The upper

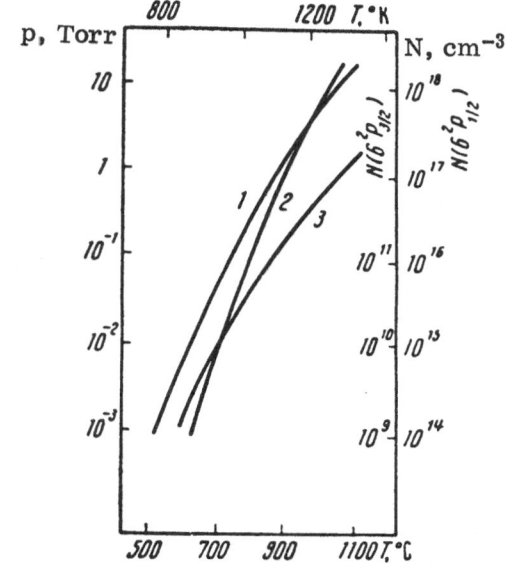

Fig. 19. Temperature dependences of the thallium vapor pressure and of the populations of the $6^2P_{3/2}$ and $6^2P_{1/2}$ levels: 1) vapor pressure; 2) population of $6^2P_{3/2}$; 3) population of $6^2P_{1/2}$.

level of this transition has a lifetime of 7.4 nsec. However, under normal conditions the radiation from the ground state is completely reabsorbed and the level lifetime is governed only by the radiative decay via the active transition. The lifetime is about 15 nsec.

The necessary thallium vapor density is established by heating to a fairly high temperature. Table 4 lists the densities of thallium atoms in the ground state $6^2P_{1/2}$ and equilibrium populations of the metastable state $6^2P_{3/2}$ as a function of temperature. Information on the thallium vapor is taken from [23]. The temperature dependences of the thallium vapor pressure and the populations of these levels are plotted in Fig. 19.

It is clear from Table 4 that the working temperature of thallium should be about 700-800°C and the population of the metastable level should rise rapidly with temperature, reaching 10^{11}-10^{12} cm^{-3} at the working temperature. This means that a certain threshold pumping rate of the upper active level must be exceeded to establish population inversion. A population of the upper level exceeding considerably the equilibrium population of the metastable level, which is also the lower active level, has to be established in a time of the order of the lifetime of the upper level. Thus, the excitation of thallium requires pulses with a steep leading edge (of the order of 10^{-8} sec).

This was the information available on thallium levels and transitions at the beginning of our investigation. Clearly, a definite conclusion on the possibility of stimulated emission could not be drawn from these data. We essentially assumed the validity of the general scheme discussed above, based on the assumption of weak electron excitation of metastable levels. Our experiments were based on these data and general considerations.

TABLE 4

T, °K	T, °C	$N(6^2P_{1/2})\cdot 10^{-16}$, cm^{-3}	$N(6^2P_{3/2})\cdot 10^{-11}$, cm^{-3}	T, °K	T, °C	$N(6^2P_{1/2})\cdot 10^{-16}$, cm^{-3}	$N(6^2P_{3/2})\cdot 10^{-11}$, cm^{-3}
873	600	0.11	0.003	1123	850	16.5	7.5
923	650	0.33	0,02	1173	900	32.5	22
973	700	1.1	0,1	1223	950	64	64
1023	750	3.2	0.53	1273	1000	111	167
1073	800	9.6	2,7				

We used a pulse cable transformer in the excitation of thallium vapor. A narrow part of a discharge tube was surrounded by a tubular heater. Thallium was placed in a side tube or in the wider central part of the discharge tube. The temperature in this part of the discharge tube was measured with a thermocouple. The windows, oriented at the Brewster angle, electrodes, and connecting tubes were outside the heated zone. Therefore, buffer gases were used in all the experiments. Superradiance power depended weakly on the nature of the buffer gas and it was highest for neon. Therefore, measurements were carried out using neon admixtures. The length of the working part of the tube was 20 cm and its internal diameter was either 2 or 3 mm. One electrode was a Kovar cylinder and the other was a tungsten electrode taken from an IFP-2000 lamp.

Superradiance at the wavelength of the green line of thallium (0.535 μ) was observed in the very first experiments. The superradiance was identified with the thallium line by a direct comparison with a thallium lamp, carried out using an ISP-51 spectrograph. A more reliable identification was made by investigating the hyperfine structure of the superradiance line, which was determined using a Fabry−Perot interferometer. It was found that under most favorable conditions this superradiance had four hyperfine structure components, associated with the presence of two thallium isotopes whose nuclear moments were $^1/_2$. The measured splitting of the superradiance line agreed with that reported in the literature [46]. A typical superradiance interferogram of the 0.535-μ line is shown in Fig. 20. Thus, our measurements indicated that the superradiance was observed, as expected, due to the $7^2S_{1/2} - 6^2P_{3/2}$ transition in the thallium atom.

In our first paper [47] we reported measurements of the superradiance characteristics in which we used apparatus whose time resolution was of the order of 10 nsec. It was later found that the duration of superradiance pulses was much less than 10 nsec. Measurements with a time resolution of 1 nsec could be carried out more recently. The shape, duration, and peak power of the superradiance pulses were measured with an FÉK-16 coaxial photocell, whose output signal was applied to the plates of an I2-7 time-interval meter. The spectral and absolute sensitivities of the photocell were supplied by the manufacturer; at the 0.535 μ wavelength, the sensitivity was 21.6 mA/W. Our measurements indicated that superradiance pulses were often complex and they frequently had two or three maxima. The peak output power was estimated for the highest maximum.

Fig. 20. Interferogram of a thallium superradiant line. The dispersion range of the interferometer was 0.18 cm^{-1}.

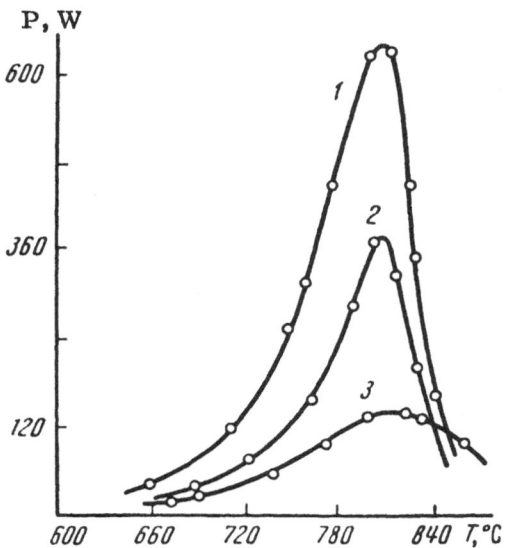

Fig. 21. Typical temperature dependences of the peak superradiance power emitted from thallium at different neon pressures (Torr): 1) 4; 2) 1.3; 3) 10. The voltage applied to the cable transformer was 15 kV.

Under the experimental conditions described above the green thallium superradiance line was observed even without mirrors. However, in the presence of one mirror behind the tube there was usually a considerable increase in the output power and directionality of the stimulated radiation. For example, without mirrors a superradiance spot at a distance of 1 m from the tube had a diameter of about 2 cm, which corresponded approximately to the geometric aperture of the tube (the tube was of 3 mm diameter and 20 cm long). When a mirror was placed at a distance of 19 cm from the working part of the tube, this superradiant spot, observed at the same distance, had a diameter of 7 mm. If was also found that an increase of the distance between the mirror and the tube improved the directionality of the superradiance but reduced its power. At some distance from the tube the mirror ceased to influence the stimulated radiation emerging from the discharge tube. Therefore, the dependences of the superradiance power on the experimental conditions were usually determined with a mirror located close to the discharge tube. The minimum distance between the mirror and the working part of the tube was usually 19 cm and could not be reduced further because of the construction of the tube. Under these conditions we obtained, inter alia, the dependences plotted in Figs. 21-23.

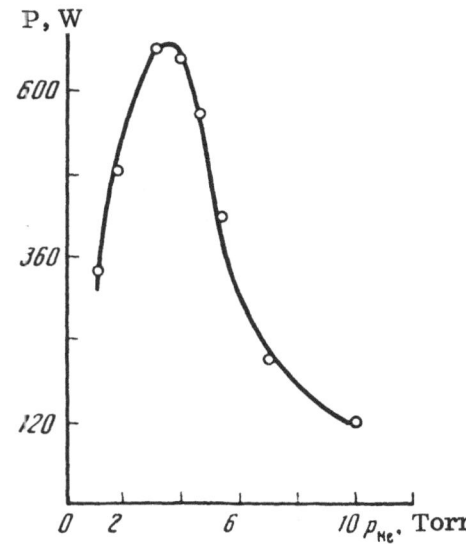

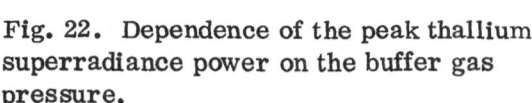

Fig. 22. Dependence of the peak thallium superradiance power on the buffer gas pressure.

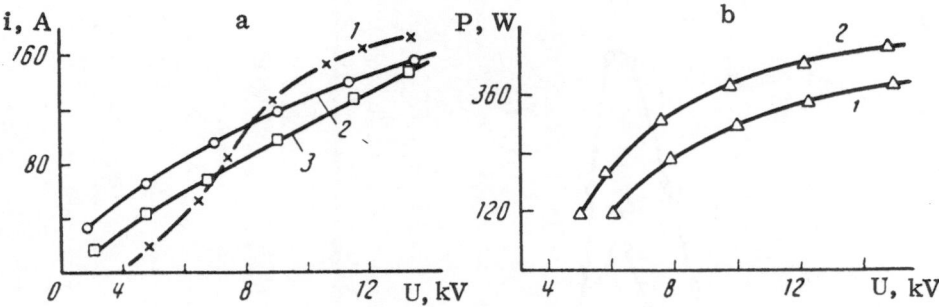

Fig. 23. Dependences of the current pulse amplitude (a) and of the peak superradiance pulse power emitted from thallium (b) on the voltage applied to the cable transformer. a: 1) p_{Ne} = 1.3 Torr, T = 780°C; 2) p_{Ne} = 4 Torr, T = 800°C; 3) p_{Ne} = 10 Torr, T = 800°C. b: 1) p_{Ne} = 1 Torr, T = 780°C; 2) p_{Ne} = 3 Torr, T = 760°C.

Under our experimental conditions the green thallium superradiance appeared beginning from about 600°C, which corresponded to a thallium vapor pressure of $7 \cdot 10^{-3}$ Torr. When the temperature was increased, the superradiance power rose, passed through a maximum, and began to fall; the stability decreased with rising temperature. At 900°C (corresponding to a thallium pressure of about 3 Torr) the superradiance disappeared. Typical temperature dependences of the peak superradiance power are plotted in Fig. 21 for three pressures of the buffer gas (neon). It should be noted that the use of a mirror behind the tube increased the output power only at some temperatures and buffer gas pressures. The presence of a mirror had a considerable influence on the output power only at temperatures up to 780°C, which corresponded to thallium vapor pressures up to 0.4 Torr. When the temperature was increased further, this mirror had no influence on the superradiance power, divergence of the radiation increased, and approximately the same power was obtained from both ends of the discharge tube. Under these conditions the stability of the superradiance pulse amplitude deteriorated. Figure 21 gives the values of the peak power for the pulses with the greatest amplitude at a given temperature. We can see that the maximum peak power corresponds to about 800°C.

Figure 22 gives the dependence of the peak superradiance power on the buffer gas (neon) pressure. This curve was obtained at 800°C for a voltage of 15 kV applied to the pulse cable transformer. The optimal neon pressure was close to 3.5 Torr. In the case of heavier inert gases the optimal pressure shifted toward lower values.

The peak superradiance power varied somewhat with the duration of operation of the thallium discharge tube. The results plotted in Figs. 21-23 were obtained for steady-state conditions established after several hours of operation at 750-800°C. These conditions were maintained for a long time provided temperature was not increased beyond 800°C. The highest superradiance power obtained in our experiments for a tube freshly filled with thallium was 900 W from each end. This corresponded to a specific peak output power of 1.3 kW/cm^3. This was the highest specific power obtained so far for pulse laser utilizing metal vapors. It was comparable with the specific laser emission power in the ultraviolet range reached in longitudinal discharges due to transitions in nitrogen molecules, which was the highest recorded specific peak power for any pulse gas laser. The cited peak power of thallium was obtained in a tube 20 cm long and of 3 mm diameter at a temperature of 800-810°C when the neon pressure was 3.5 Torr and the voltage applied to the cable transformer was 15 kV.

Figure 23 gives typical dependences of the peak superradiance power on the voltage applied to the primary winding of the transformer. Under all conditions there was a tendency for the peak power to saturate with rising voltage. This figure includes also the dependences of the current pulse amplitude on the voltage applied to the transformer. The current amplitude also tended to saturation.

Figure 24 shows oscillograms demonstrating how the superradiance power, shape, and duration of pulses at 0.535 μ wavelength varied with temperature at different buffer gas (neon) pressures. The attenuation of the light beam reaching a coaxial photocell is given on the left of each curve; the temperature in the discharge tube is given on the right. These oscillograms were obtained using an I2-7 oscillograph when the voltage applied to the transformer was 15 kV and the mirror was placed at a distance of 19 cm from the working part of the tube. The superradiance was collected on the photocell photocathode by a lens. The oscillograms indicated that the pulse duration at midamplitude did not exceed 3 nsec. In the temperature range 600–780°C the superradiance pulses had a complex shape with two or even three maxima. When the temperature was increased, the pulse shape became simpler and it just had one maximum. A similar change in the pulse shape occurred when the neon pressure was

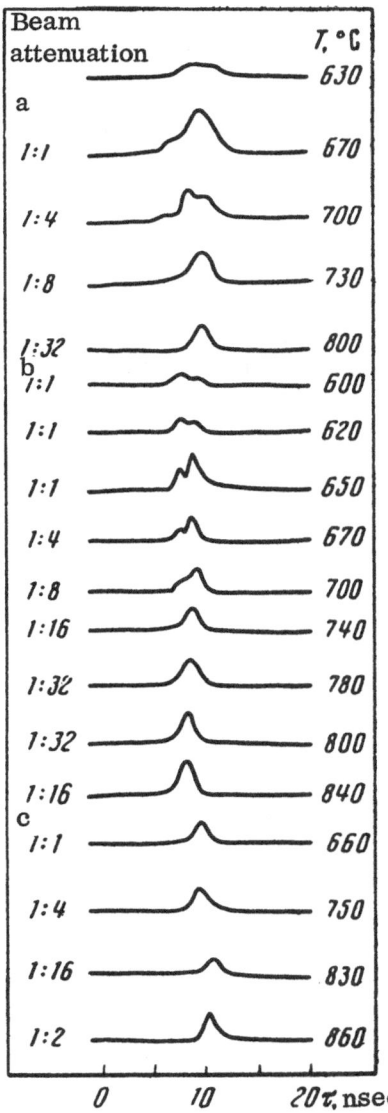

Fig. 24. Influence of the temperature in the discharge tube on oscillograms of thallium superradiance pulses recorded at three neon pressures (Torr): a) 1.3; b) 4; c) 10.

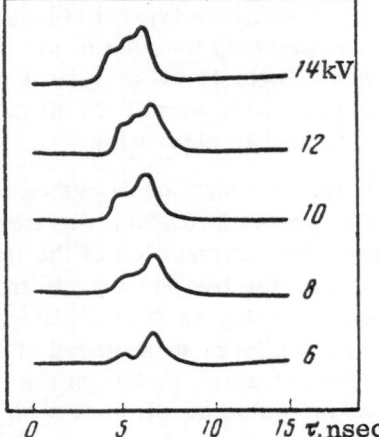

Fig. 25. Influence of the voltage applied to the cable transformer on oscillograms of thallium superradiance pulses recorded at a neon pressure of 4 Torr, T = 700°C.

increased. Above 8 Torr the superradiance pulses had only one maximum at all temperatures. This increase in temperature and buffer gas pressure not only altered the shape of the super-radiance pulses but also their duration. At 600-700°C the pulse duration was about 3 nsec, whereas at 820-870°C it was 0.5 nsec. When the neon pressure was 10 Torr, the pulse duration did not exceed 1.5 nsec at all temperatures.

Figure 25 shows oscillograms of superradiance pulses obtained for different voltages applied to the transformer. These voltages ranged from 6 to 14 kV and the temperature was 700°C. The neon pressure was 4 Torr. It is clear from Fig. 25 that the shape of the super-radiance pulses changed when the voltage was increased.

We also investigated in greater detail the influence of a mirror behind the tube on the shape and duration of the thallium superradiance pulses. Figure 26 shows oscillograms of superradiance pulses obtained in the presence and absence of a mirror under various experimental conditions. In practice, oscillogram 2 was recorded with a mirror covered with a black paper. The distance from the mirror to the working part of the tube was 19 cm. It is clear from Fig. 26 that, under certain conditions, the presence of a mirror had a considerable influence on the shape of the suerradiance pulses. The mirror had practically no effect on the initial part of the pulse but an additional power peak frequently appeared in the final part. The amplitude and shape of this peak depended on the experimental conditions and its position on

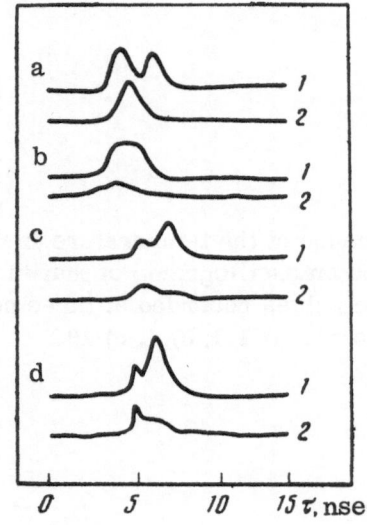

Fig. 26. Oscillograms of thallium superradiance pulses obtained in the presence (curves denoted by 1) and absence (curves denoted by 2) of a mirror behind the discharge tube: a) T = 660°C, p_{Ne} = 6.7 Torr; b) T = 730°C, p_{He} = 4 Torr; c) 700°C, p_{He} = 7.6 Torr; d) T = 730°C, p_{Ne} = 6.7 Torr.

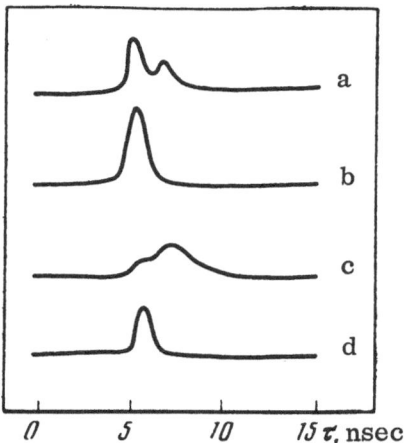

Fig. 27. Influence of the buffer gas temperature and pressure on the shape of superradiance pulses: a) $T = 660°C$, $p_{Ne} = 6.7$ Torr; b) $T = 720°C$, $p_{Ne} = 6.7$ Torr; c) $T = 720°C$, $p_{He} = 6$ Torr; d) $T = 870°C$, $p_{He} = 6$ Torr.

the time axis was governed by the distance between the mirror and the discharge tube. The presence of a mirror increased considerably the amplitude of this peak. The appearance of the second peak was attributed to light amplified in the active medium, reflected from the mirror, and passed for the second time through the active part of the tube and, therefore, subject to some time delay. The addition of a second mirror, i.e., formation of a resonator, usually did not alter greatly the shape of the superradiance pulses. The influence of the resonator was manifested simply by the appearance of an additional small maximum at the end of the superradiance pulse. This was due to the short inversion lifetime. Therefore, there was no point in investigating the influence of a resonator with a high Q factor.

As mentioned earlier, the influence of a mirror on the superradiance pulses was observed only under certain conditions. These conditions were low temperatures and low pressures of the buffer gas. At high temperatures and pressures the presence of a mirror had no influence on superradiance. In the latter case the superradiance pulses assumed a simple shape and their duration decreased. Figure 27 demonstrates the influence of temperature on pulse duration. It is clear from this figure that the duration of superradiance pulses decreased by a factor of about 3 when the temperature in the discharge tube was increased and it amounted to 1 nsec at 870-900°C. Since the time resolution of our apparatus was of the same order of magnitude, the true duration of the pulses was probably even less.

We shall now consider the nature of changes in the superradiance pulses as a function of the distance between the mirror and the working part of the tube. It is clear from Fig. 26 that when a mirror was placed behind the tube, a second peak appeared (under certain conditions) and this peak was delayed relative to the first. Measurements indicated that this delay represented the time taken by a light pulse to travel from the working part of the tube to the mirror and back again. The results plotted in Fig. 26 were obtained when the distance from the mirror to the working part of the tube was 19 cm and the delay was 1.4 nsec. An increase in the distance from the mirror to the tube caused the additional power peak to shift further along the pulse and to decrease in amplitude until it disappeared completely.

Since the directionality of the radiation reflected from the mirror and amplified by the second passage through the active medium was considerably better than in the absence of a mirror, it was possible to study the shape of the pulses of the amplified superradiance in a separate experiment. In this case we recorded superradiance without a lens in such a way that the bright spot produced by the mirror covered completely the photocathode of the coaxial photocell. In the absence of a mirror the size of the superradiant spot was considerably greater than the photocathode area so that practically no signal was recorded. Oscillograms of amplified superradiance pulses were obtained (Fig. 28) for different voltages on the trans-

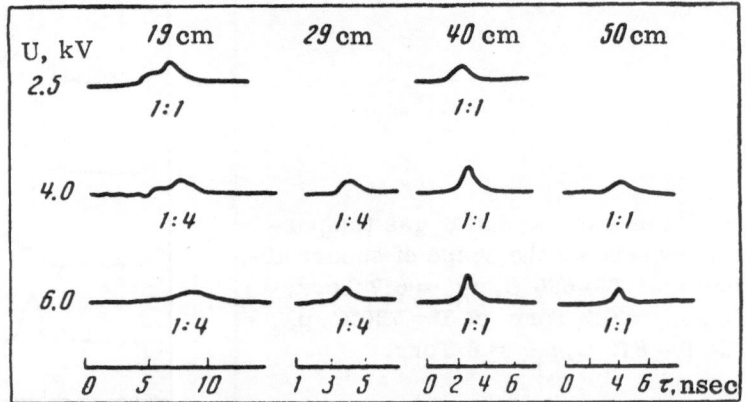

Fig. 28. Influence of the discharge between the mirror and the active medium (given at the top of the figure) on the oscillograms of superradiance pulses amplified by the return pass through the active medium. The voltage applied to the cable transformer is given on the left-hand side of the figure and the attenuation of light before reaching a coaxial photocell is given below each curve.

former, as a function of the distance between the mirror and the working part of the tube. The temperature in the tube was 700°C and the neon pressure was 4 Torr. The attenuation of superradiance reaching the photocell was varied (the actual attenuation is given under each curve). When the distance from the mirror to the working part of the tube was increased, the superradiance pulses became simpler and shorter. It should be stressed that very short pulses were obtained for large distances between the mirror and the tube. For example, when this distance was 40-50 cm, pulses of about 1 nsec duration at the base were recorded. This measured duration was probably governed by the resolution of the apparatus and the true duration was even less.

Figure 29 shows the relative positions of the current and stimulated radiation pulses. A current pulse and a superradiance pulse were applied simultaneously to the input of I2-7 through a mixer. The times taken by the signals to travel along the connecting cables were measured and selected to be the same to within 1 nsec. An estimated delay in the coaxial

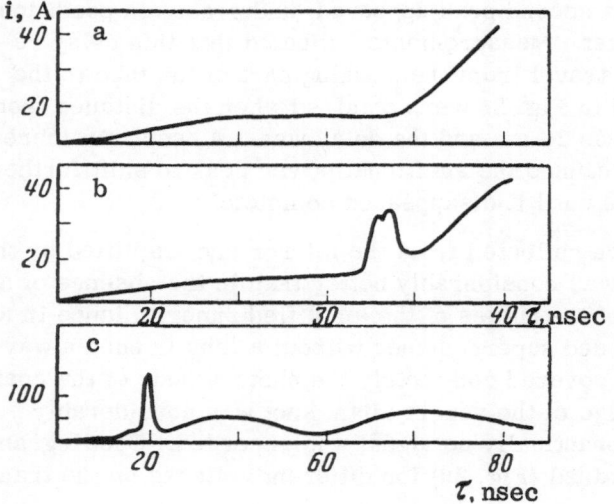

Fig. 29. Position, on the time axis, of superradiance pulses relative to the current pulse: a) initial part of the current pulse; b) current and superradiance pulses; c) current and superradiance pulses on a different scale.

photocell was short. An oscillogram obtained by recording the current alone is shown in Fig. 29a whereas Fig. 29b is the result of simultaneous recording of the current and superradiance. Clearly, superradiance appeared at the very edge of the current pulse and its duration was considerably less than the current rise time. The relative durations of the current and superradiance pulses can be seen more clearly in Fig. 29c, which is an oscillogram on a different time scale, obtained for a somewhat weaker current pulse. These oscillograms were recorded when the temperature in the tube was T = 800°C, the neon pressure was P_{Ne} = 4 Torr, and the voltage applied to the transformer was 15 kV; i.e., they were recorded under conditions which were near optimal from the point of view of the highest peak power. It is clear from the oscillograms that under our conditions the excitation (current) pulse was very poorly matched to the superradiance pulse and a very considerable part of the excitation energy was lost uselessly. Clearly, under these conditions the efficiency could not be high.

We carried out a preliminary study of the influence of pulse repetition frequency on the properties of superradiance. We discharged a 2200-pF capacitor through a TGI1-500/16 thyratron and the primary winding of the pulse cable transformer. We used trains of regular pulses as well as pulse pairs. In the latter case the thyratron was triggered by two pulses which could be separated by a variable delay ranging from 300 to 2000 μsec. The frequency of these pulse pairs was 2–10 Hz. The two current pulses were practically identical in shape and amplitude for delays between 2000 and 550 μsec. In the delay range 500–300 μsec, the amplitude of the second pulse decreased compared with the first.

Our measurements indicated that when the delay was varied within the range 2000–500 μsec, the power of the second superradiance pulse emitted from a tube freshly filled with thallium remained constant and was approximately half the power of the first pulse. When the delay time was 500–300 μsec, the power of the second pulse decreased and this was clearly due to a reduction in the current pulse amplitude.

When a thallium-filled tube was operated for a long time, the peak power of the second pulse exceeded the power of the first one. In some cases we even observed a situation in which the first excitation pulse produced no superradiance, whereas the second generated superradiance of considerable power (100 W). The shape of the second superradiance pulse depended on the experimental conditions and sometimes differed considerably from the shape of the first pulse.

The application of a train of regular pulses demonstrated that the superradinace power decreased when the pulse repetition frequency was increased. Beginning from about 200 Hz the superradiance power was not greatly affected by the frequency.

It was worth noting particularly that, beginning from a repetition frequency of 200 Hz, the superradiance pulses became extremely stable in respect of their amplitude and the time of appearance relative to the excitation pulse. This was also observed for pulse pairs. The superradiance pulse produced by the second excitation pulse was, in contrast to the first pulse, more stable in respect of its amplitude and time of appearance. The scatter of the time of appearance relative to the excitation pulse did not exceed 0.1 nsec. This indicated that when the time interval from the preceding pulse was short (of the order of 5 nsec), the breakdown and discharge conditions became very stable for the next pulse.

These preliminary investigations did not give a full picture of the behavior of the 0.535 μ superradiance when the repetition frequency was increased. Further experiments would be needed, particularly with a shorter delay between pulses in pairs. However, it was clear that the green pulse superradiance of thallium with a high peak power could be obtained at repetition frequencies up to 2–3 kHz and possibly even higher. A very important observation was that the stability of superradiance pulses in respect of their ampitude and time of appearance improved strongly when the pulse repetition frequency was increased.

Information on the true duration of superradiance pulses and the appearance of these pulses relative to the current pulses made it possible to estimate the efficiency of conversion of the discharge energy during a superradiance pulse. However, only a rough estimate was possible because the precision of measurements of the parameters of short current pulses was low. A rough estimate based on the ratio of the pump power during superradiance to the super-radiance power indicated that under maximum peak power conditions the energy conversion efficiency was of the order of 0.5-0.2%. At lower voltages the energy conversion efficiency was of the order of 1%. It depended also on the nature of the thyratron employed in the power supply circuit. Estimates indicated that when the TGI1-130/10 thyratron was used, the energy conversion efficiency could, under certain conditions, reach 1-2% during the emission of superradiance.

Under our conditions a considerable proportion of the energy was lost in wide connecting tubes of about 10 mm diameter, which joined the electrodes outside the hot zone to the narrow heated part of the tube. The thallium pressure in the wide cold tubes was negligible and the discharge energy was lost in the excitation and ionization of the buffer gas. The energy losses in these parts of the discharge tube were difficult to estimate. Moreover, there was a gradient of the thallium vapor density along the working part of the tube, so that part of the active medium was known to be under nonoptimal conditions. These factors reduced the energy conversion efficiency and the peak superradiance power. The results obtained indicated that the efficiency of conversion of the excitation energy into stimulated radiation emitted as a result of the investigated thallium transition could reach 5% under conditions optimal from the point of view of efficiency and the specific peak power could reach 10 kW/cm^3.

In Chap. II we gave a calculation of the saturated stimulated radiation power at the wavelength of the 5350 Å line. The results of this calculation were in qualitative agreement with the experimental data. The calculation was carried out for a three-level system subject to some simplifying assumptions. However, the active medium used in our experiments was not homogeneous and it was difficult to say to what extent the measured superradiance power was close to saturation. In view of this, there was no point in making quantitative comparisons of calculations with experiment. However, according to the calculations a high stimulated radiation power should be reached only at electron densities such that one would have to allow for collisional transitions between the active levels. The duration of superradiance pulses found experimentally was, in agreement with the calculations, considerably shorter than the lifetime of the upper active level but still significantly less than predicted by the calculations. The cause of this discrepancy was not clear.

The results reported above could be used to draw certain conclusions on the factors responsbile for the low efficiency of a thallium vapor laser. The main factor was the mismatch between the radiation and excitation (current) pulses. Clearly, a higher efficiency would be obtained by the use of current pulses lasting only a few nanoseconds. Such pulses could be obtained if the whole discharge circuit, including the discharge cavity, had a very low inductance. This would impose certain restrictions on the choice of the discharge cavity configuration. The power supply system could be improved in various ways. These ways are discussed briefly in the conclusions. Here, we shall point out that the special difficulty in the construction of a fast supply system for a thallium vapor laser is related to the need to operate at a high temperature. The efficiency can be increased by constructing the discharge cavity in such a way that the discharge is concentrated in the active zone containing thallium and the same temperature is maintained throughout the active zone. In this case the temperature might be reduced somewhat since under diffusion flow conditions it is necessary to heat a tube to a higher temperature.

Thus, it would be desirable to construct sealed discharge cavities with windows and electrodes located in a zone where the temperature is 700-800°C. The best configuration would be a cavity suitable for transverse discharges because it is then possible to construct a low-inductance high-power system. The development of such a system and discharge cavity meets with technical difficulties. It is at present impossible to say to what extent and how these difficulties may be overcome. If they are overcome, one may expect specific output powers of the order of 10 kW/cm^3. This means that an active medium of moderate volume (~ 10 cm^3) can give a peak output power up to 100 kW. Moreover, if the duration of the excitation pulses is made of the order of the duration of the superradiance pulses, the efficiency should be relatively high amounting to several percent. In practice, one would have to consider the loss of energy due to the heating of the cavity. This loss can, in principle, be minimized by ensuring good thermal insulation and compensating the losses due to heat conduction and radiation. Such losses can be compensated at least partly by the heating caused by the discharge, particularly at high pulse repetition frequencies.

The results obtained at high repetition frequencies for pulse pairs suggest that when tubes have been operated for a long time a considerable number of thallium atoms is detached from the tube walls by a discharge. Thus, at a high repetition frequency one could reduce the temperature of the outer walls of the tube. Further experiments would be needed before establishing the conditions suitable for operation at high repetition frequencies.

§ 3. Pulse Superradiance due to Green Line of Thallium in TlI Vapor

Pulse superradiance due to the 5350 Å atomic thallium line was observed by us in TlI vapor [48] during a search for methods of introducing thallium atoms into a discharge tube with walls kept at a relatively low temperature. This superradiance was also of interest because of a new population inversion mechanism involving selective population of the one of the atomic levels during dissociation of a molecule by electron impact.

This mechanism may be of interest in the search for high-efficiency systems because, in principle, η_{ult} can be fairly high. However, it is not yet possible to estimate η_{ult} for this mechanism because no information is available on the electron-impact dissociation of molecules.

Our apparatus did not differ geatly from that described in Chap. II. Quartz discharge tubes had a working length of 200 mm and their internal diameter was 1.3 mm. Thallium iodide was placed in a side tube or in the wider central part of the tube. The temperature was measured in the direct vicinity of the side tube. The temperature dependence of the vapor pressure of thallium iodide was taken from [49]. In some cases the whole tube with windows and electrodes was placed inside a heater, whereas in other cases only the working part of the tube was so enclosed. Inert buffer gases were used. Some experiments were carried out in a sealed tube containing TlI vapor without buffer gases. A pulse discharge in the tube was excited with the cable transformer described earlier. The repetition frequency was varied from a few hertz to 1 kHz. No resonator was used. An aluminized mirror was usually placed behind the discharge tube and the radiation emitted from the tube end was focused onto the slit of a DFS-12 spectrograph. The mirror and the whole system were aligned using the 6143 Å neon superradiance line. The visible and infrared spectra were recorded using FÉU-36 and FÉU-28 photomultipliers, respectively. The photomultiplier signal was applied to the input of the amplifier of an S1-11 oscillograph. The time resolution of the recording system was 7-10 nsec. The wavelength was measured to within an estimated error of 0.5 Å.

When the whole discharge tube, together with the electrodes and windows, was placed inside the heater, the 5350 Å superradiance line was observed in the temperature range 370-440°C. Helium, neon, xenon, and argon were used as the buffer gases. The superradiance power increased in the sequence helium, neon, argon, and xenon approximately in the ratio

1 : 10 : 20 : 30. The optimal buffer gas pressure was 10 Torr for helium, 6 Torr for neon, 1.3 Torr for argon, and 0.7 Torr for xenon. Experiments were also carried out in pure thallium iodide vapor without a buffer gas. In this case the tube containing thallium iodide was carefully out-gassed at the working temperature and then sealed at a residual gas pressure of 10^{-5} Torr. The 5350 Å superradiance line was emitted by pure thallium iodide vapor in the sealed tube when the temperature was 370-390°C. A mirror placed behind the tube had no significant influence on the superradiance power. This experiment then demonstrated that the buffer gas was not essential for superradiance. However, the superradiance power was considerably lower in the absence of such a gas.

When the whole tube was heated, we encountered additional difficulties associated with the damage caused to the electrodes and with the breakdown between the electrodes and the heater. Therefore, most of the experiments were carried out with only the working part of the tube inside the heater. In this case the thallium superradiance was observed for all the buffer gases mentioned above in the temperature range 430-490°C and the maximum power for all these gases was achieved at the same temperature (to within 10°C). The range of temperatures in which superradiance was observed and the temperature corresponding to the maximum power differed from the values obtained when the whole tube was heated. Considering all factors, we concluded that the difference was due to the fact that when the central part of the tube was heated a diffusion current and a gradient of the active substance along the tube were established. Therefore, a given thallium iodide vapor density could be reached only if the central part of the tube was heated to higher temperatures than in the case when the whole tube was heated. From these results we concluded that the maximum superradiance power was reached at a definite thallium iodide density, which was ~5 · 10^{15} cm³, which corresponded to 380°C. The dependences of the superradiance power on the buffer gas pressure are plotted for this temperature in Fig. 30. The range of buffer gas pressures in which superradiance was observed varied from gas to gas, becoming narrower on transition from neon to xenon. The superradiance power maximum was reached at 6 Torr for neon, 1.3 Torr for argon, and 0.7 Torr for xenon. The superradiance power maxima for xenon, argon, and neon were approximately in the ratio 3 : 2 : 1. Once again the placing of a mirror behind the tube had little effect on the superradiance power irrespective of the experimental conditions.

It was interesting to note that when the buffer gas pressure was suitably selected, it was possible to observe simultaneously superradiant lines of the buffer gas (6143 Å for neon, and 9799 Å and 9045 Å for xenon) and of thallium (5350 Å). The superradiance power of the buffer

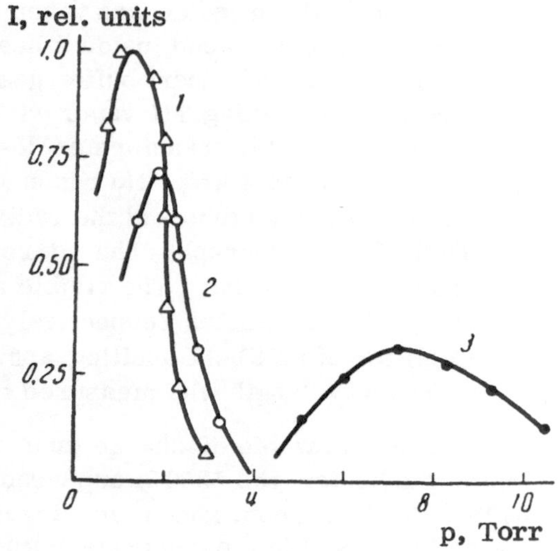

Fig. 30. Dependences of thallium superradiance power, emitted from thallium iodide vapor, on the buffer gas pressure at T = 460°C: 1) xenon; 2) argon; 3) neon.

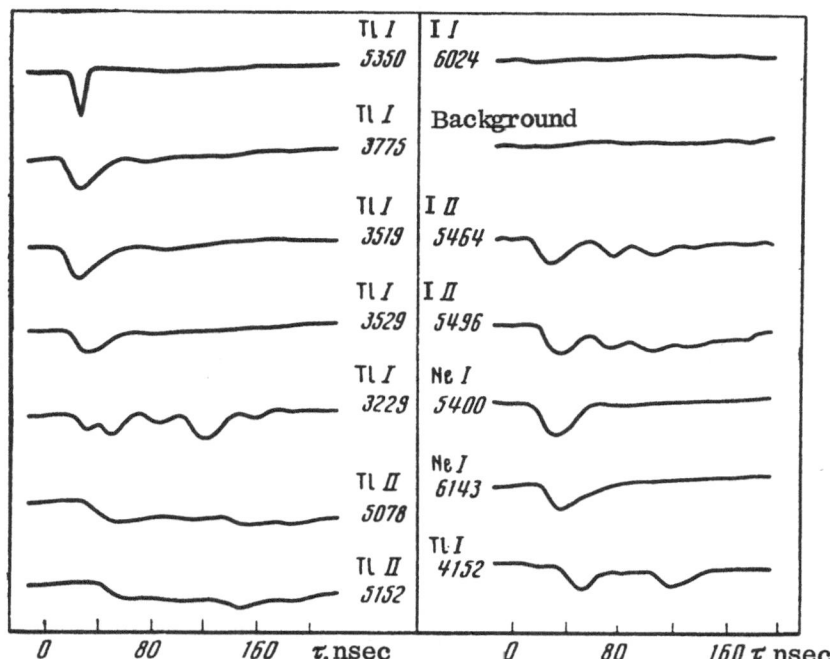

Fig. 31. Oscillograms of 5350 Å superradiance pulses and of spontaneous radiation lines. The time scale is 40 nsec/div.

gases decreased with rising temperature and at high temperatures it disappeared completely. For example, when the neon pressure was 1.5 Torr, the 6143 Å neon line superradiance was observed in the temperature range 20-380°C. Under the same conditions the 5350 Å thallium line superradiance was emitted in the range 370-440°C.

The spectra and time characteristics of the spontaneous and stimulated radiation emitted from our discharge tubes were investigated photoelectrically in the spectral range 3000-6000 Å and in a temperature range 360-520°C, i.e., in a temperature range somewhat wider than the existence of the 5350 Å line superradiance. Under these conditions we observed lines due to I II, Tl I, Tl II, and Tl III. We also observed a line at 4152 Å, which was attributed tentatively to the TlI molecule [50]. The atomic (I) lines of iodine were not observed. At the wavelengths corresponding to the strong I I line the signal did not exceed the background level (Fig. 31). We investigated in greater detail the spectrum of a discharge with neon. We determined the time characteristics of the following spontaneous lines: 5946 and 5465 Å of I II: 3229, 3529. 3519, 3755 Å of Tl I; 5142 and 5073 Å of Tl II. Only two neon lines (6143 and 5400 Å) were investigated. The selected lines were the strongest under our experimental conditions. Consequently, they were easily identified and measured. Moreover, these lines did not participate in the postulated photodissociation of the TlI molecule, i.e., these lines could be used to characterize the discharge itself.

Among the thallium lines (Fig. 34), the resonance line at 3775 Å (transition $7^2S_{1/2}-6^2P_{1/2}$) started from the same level as the 5350 Å line. We used the former to determine the population of the $7^2S_{1/2}$ level. The 3529 Å (transition $6^2D_{3/2}-6^2P_{3/2}$) and 3519 Å (transition $6^2D_{3/2}-6^2P_{3/2}$) lines started from the $6^2D_{3/2}$ (resonance) and $6^2D_{5/2}$ lines. They terminated at the same metastable level as the 5350 Å line. We could not exclude the possibility of inversion of the levels responsible for these lines. Finally, the 3229 Å line was due to a transition from a higher level $8^2S_{1/2}$.

The shape and duration of the spontaneous radiation pulses representing the Tl lines at 3775, 3529, and 3519 Å remained constant in the investigated temperature range. The duration of the spontaneous radiation at the lines due to Tl II and $I II$ increased and their shape changed with rising temperature. Figure 31 gives oscillograms of these superradiance lines at 440°C. The neon gas pressure was 10 Torr and the rate of scan was 20 nsec/cm.

It is clear from the oscillograms in Fig. 31 that all the spontaneous lines started to emit simultaneously, within the limits of the experimental error. A superradiance pulse at the 5350 Å wavelength appeared at the leading edge of the spontaneous line and its duration was much less than that of the spontaneous radiation. The duration of the superradiance pulses in these oscillograms was 5–7 nsec at midamplitude. The use of a special photomultiplier and of the I2-7 oscillograph indicated that the true duration of the superradiance pulses did not exceed 3 nsec at midamplitude. Moreover, superradiance due to transitions in neon and xenon appeared simultaneously (within the limits of the experimental error) with the superradiance of the green thallium line. In determining the population inversion mechanism it was interesting to compare the dependences of the superradiance power and spontaneous line intensities on the experimental conditions. The superradiance power was found to vary much more strongly with the experimental conditions than the spontaneous line intensities. An analysis of the oscillograms obtained indicated that within the temperature range including the whole interval of existence of the 5350 Å superradiance line the spontaneous line intensities first rose with temperature but then the rise slowed down. The intensities of all the iodine and thallium spontaneous radiation lines increased first with the discharge temperature; this was followed by a plateau and a fall beginning from 510°C.

The shape of the spontaneous radiation pulses representing ionic lines was more complex than that of the atomic lines; these pulses had two maxima and the ratio of the amplitudes of these maxima varied with temperature. Superradiance coincided with the first maximum. When the discharge temperature was increased, the first maximum decreased relative to the second. In the temperature range where superradiance was observed at the 5350 Å wavelength, the intensities of the ionic lines did not vary greatly. Typical dependences of the intensities of the 5464 Å spontaneous line of $I II$ and of the superradiance power of the 5350 Å line of Tl I on the discharge temperature are plotted in Fig. 32.

We investigated also the intensities of the 5496 and 5464 Å spontaneous lines of $I II$ as a function of the nature of the buffer gas. The results obtained are plotted in Fig. 33 alongside the power of the 5350 Å superradiant line. We can see from this figure that a change of the

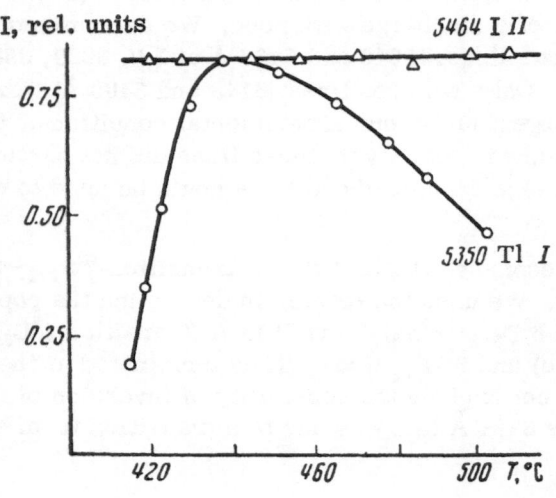

Fig. 32. Temperature dependences of the superradiance power and intensities of spontaneous radiation lines of $I II$ (p_{Ne} = 10 Torr).

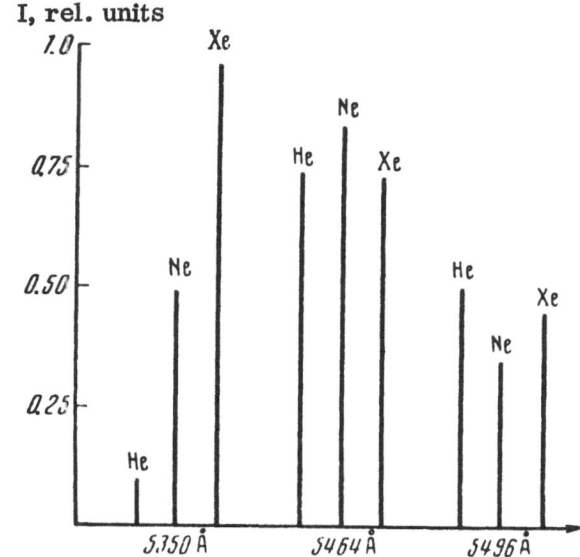

Fig. 33. Relative powers of superradiance and spontaneous radiation lines obtained at T = 480°C and different buffer pressures: p_{He} = 6 Torr, p_{Ne} = 6 Torr, and p_{Xe} = 0.66 Torr.

nature of the buffer gas could alter very greatly the superradiance power but had little effect on the spontaneous radiation intensity.

In addition to an investigation of superradiance of thallium iodide vapor, we decided to study superradiance using the analogous molecule of thallium bromide. The experiments were carried out using the same apparatus and the same discharge tube. The buffer gases were neon, helium, argon, and xenon. Their pressures ranged from 0.5 to 6 Torr. The tube was heated to 520°C, which corresponded to the thallium bromide pressures up to 10 Torr. The emission spectrum was recorded photoelectrically in the 3000-6500 Å range. The strongest lines were due to the thallium atom (3775 and 5350 Å). No superradiance lines were observed in a wide range of experimental conditions. This absence of superradiance in the case of thallium bromide vapor could be due to the different potential energy curves of the thallium iodide and bromide molecules. In particular, the metastable level of thallium could be populated in a different way.

The potential curves of the thallium iodide molecule [50, 51] and the energy level scheme of the thallium atom are given in Fig. 34. We have to consider various ways of establishing a population inversion of the thallium atomic transition in question. We may assume that a discharge tube contains not only thallium iodide but also metallic thallium and that stimulated emission occurs in thallium vapor. Superradiance and laser emission from thallium vapor are known to occur at the same 5350 Å line (see § 2). In the present experiments no thallium was added to the discharge tube. However, it could be formed as a result of decomposition of the TlI molecules in the discharge. Nevertheless, a significant thallium vapor pressure could not be produced at the relatively low temperatures used in experiments. Superradiance due to thallium vapor was observed beginning from 600°C. At the working temperatures in the present experiments the thallium vapor pressure was negligible. Thallium vapor could appear also in the discharge as a result of thermal decomposition of the TlI in the hotter central part of the discharge.

This mechanism of emission of the atomic thallium lines was suggested in [5] for gas-discharge lamps containing iodide admixtures, particularly thallium iodide. It was assumed that the temperature of the discharge wall tubes corresponded to a Tl I vapor pressure of about 1 Torr and the temperature on the axis of the discharge tube in the central zone was so much higher that thermal decomposition of the molecules into atoms took place on the axis.

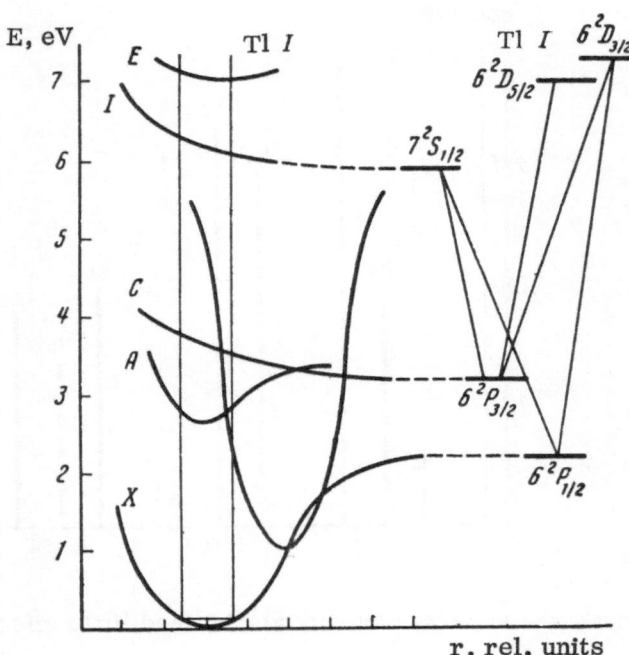

Fig. 34. Potential energy curves of the Tl I molecule and energy-level and transition scheme of the thallium atom Tl I.

It was also assumed that thallium atoms were not deposited on the tube walls because they recombined again with the iodine atoms in the course of their diffusion to the walls. Under our experimental conditions this mechanism could not occur. In a tube with a small diameter operated at a low pulse repetition frequency the temperature gradient along the diameter should be negligible. Consequently, the thermal decomposition of thallium iodide could not take place at the working temperatures in our tubes. This was supported by the absence of the atomic iodine lines from the discharge spectrum. The excitation potentials of the iodine and thallium lines were fairly close so that when molecules dissociated into thallium and iodine in the ground states, we should observe not only thallium but also iodine atomic lines.

We had to assume that the thallium atom levels were populated as a result of dissociation of thallium iodide molecules from excited electron states. In principle, these molecules could also be photodissociated by the radiation generated in the discharge and they could also be split by electron impact.

Let us consider the possibility of photodissociation resulting in the formation of thallium atoms in the $7^2S_{1/2}$ state. The photodissociation to this state gives rise to an absorption band of the Tl I molecule extending from about 1900 to 2100 Å with a maximum at 2000 Å [52]. In this spectral range the lines which can be emitted from a discharge in thallium iodide vapor, including a discharge in the presence of a buffer gas, may be several strong lines of I I, I II, and Ar III. Experiments with a sealed tube indicated that the presence of a buffer gas was not necessary for superradiance of the green thallium line. Therefore, we should consider the I I and I II lines. As mentioned earlier, the I I lines were not observed at all. Therefore, it was unlikely that they could appear in the region near 2000 Å. The I II lines were observed easily under our experimental conditions. However, they could be hardly responsible for the excitation of the 5350-Å line because their behavior was not correlated with the behavior of the super-radiance and spontaneous radiation lines starting from the $7^2S_{1/2}$ level. It is clear from the oscillograms in Fig. 31 that the duration of the spontaneous radiation emitted from this level was considerably less than the duration of emission of the I II lines. Moreover, when the temperature was varied, the superradiance power varied quite rapidly (Fig. 32), whereas the intensity of the I II lines was hardly affected.

When buffer gases were used, the superradiance power increased. This could not be attributed to photodissociation. In particular, the power depended strongly on the nature of the buffer gas and it was approximately twice as high for xenon as for argon. Moreover, no strong xenon lines were observed in the 2000 Å range, whereas they were emitted by argon. In the case of the I II lines the intensity was not greatly affected by the nature of the buffer gas. Thus, we concluded that although the photodissociation mechanism could, in principle, give rise to a population inversion of the 5350 Å lines, this did not occur under the conditions in our experiments because of the absence of strong lines in the relevant part of the spectrum.

Thus, we were left with the possibility that the $7^2S_{1/2}$ thallium levels were populated, in the course of dissociation of the thallium iodide molecules, by electron impact. This mechanism was in good agreement with all the observations. In fact, it is clear from the positions of the potential curves and from the Franck–Condon principle that the dissociation of the TlI molecule by electron impact or photodissociation should produce thallium atoms in the $7^2S_{1/2}$ state [53]. The dependence of the superradiance power on the experimental conditions and on the nature of the buffer gas can easily be explained by changes in the conditions in a plasma, particularly changes in the electron temperature.

Thus, an analysis of our results yields the conclusion that the main population inversion mechanism of the 5350-Å thallium line in thallium iodide vapor discharges is, irrespective of the presence of buffer gases, the dissociation of the thallium iodide molecules by electrons accompanied by a preferential population of the $7^2S_{1/2}$ level.

§ 4. Laser Emission and Superradiance due to

Transitions in Lead

Pulse laser emission from lead vapor was first observed for the $7s\,^3P_1^0 - 6p^2\,^1D_2$ (7229 Å) transition in 1965 [9]. This was the first experimental detection of laser action due to a transition from a resonance to a metastable level. Subsequently, a record gain of 600 dB/m and a peak output power of 2 kW were obtained for this transition [54]. This power was generated in a tube of 10 mm diameter and 10 cm long excited by current pulses with a leading edge of 150 nsec and amplitude of 1500 A. The isotopic splitting of the 7229 Å line was observed under superradiance conditions in [55]. The ultimate efficiency of the 7227 Å line was $\eta_{ult} = 0.24$. It was somewhat lower than for other lines due to transitions from resonance to metastable levels, for example, thallium or copper lines.

We were interested in laser emission from lead vapor because of other atomic transitions in lead from resonance to metastable levels characterized by a high ultimate efficiency. Attention to this point was paid also in the analysis of possible high-efficiency systems.

The energy level and transition scheme of the lead atom is shown in Fig. 35. The lead atom has two resonance levels $6p7s^3P_1^0$ and $6p6d^3D_2^0$ coupled to the ground state by strong resonance lines at 2833 and 2170 Å. The lead configuration $6p^2$ describes not only the ground state but also four metastable levels: $^1S_0$, $^1D_2$, $^3P_1$, and $^3P_2$. Thus, there are several transitions between resonance and metastable levels. In particular, the $6p7s^3P_1^0$ level, from which the 7229 Å line begins, can give rise also to two transitions to the $6p^2\,^3P_{1,2}$ levels with wavelengths 3639 and 4057 Å. If we use the general population inversion mechanism discussed above and assume that the effective excitation cross sections of the metastable levels are much smaller than the corresponding cross sections of the resonance levels, we can expect pulse inversion for these lines. However, although stimulated emission from lead has been investigated many times, only the 7229 Å line was observed.

We shall try to explain this situation by estimating the gain for the transitions under consideration. The gain of a Doppler-broadened line is proportional to the product of λ^3A_{ik}

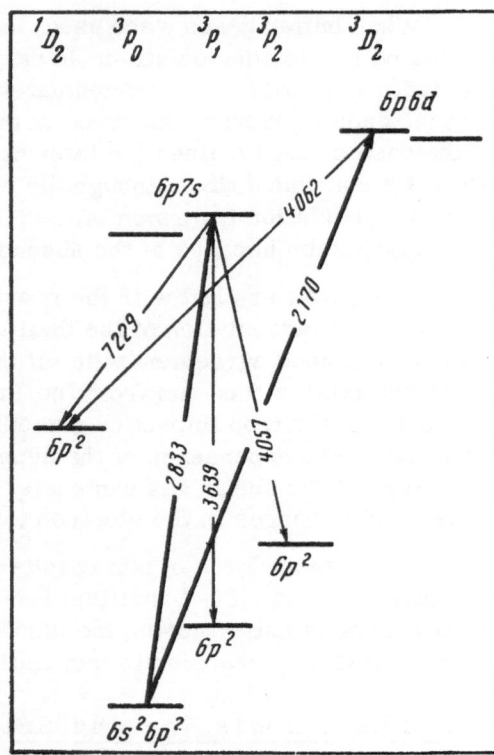

Fig. 35. Energy-level and transition scheme of the lead atom.

and of the population inversion, where A_{ik} is the probability of the active transition. If we assume, in accordance with the general mechanism, that during the initial moments of an excitation pulse the metastable level population is negligible, we find that the gain of the lines beginning from the same level is proportional to $\lambda^3 A_{ik}$. Table 5 lists the characteristics of the transitions of interest to us, including the products $\lambda^3 A_{ik}$. In this table f_{ik} denotes the oscillator strength and $\tau_{ik} = 1/A_{ik}$. Information on the oscillator strengths and transition probabilities is taken from [56], and for the lines identified by an asterisk, approximate data are taken from [57].

It is clear from Table 5 that in the case of a weak population of the metastable levels, among the lines beginning from the $6p7s\,^3P_1^0$ level the highest gain should be exhibited by the 4057 Å line, and the 3639 and 7229 Å lines should have comparable gains. The question is now why laser emission or superradiance is not observed at the 4057 and 3639 Å wavelengths. Clearly, these lines are not observed because our assumption of low populations of the meta-stable levels is incorrect, i.e., the general population inversion mechanism described above

TABLE 5

λ, A	Transition	f_{ik}	$A_{ik} \cdot 10^8$, sec^{-1}	τ_{ik}, nsec	$\lambda^3 A_{ik}$	$\dfrac{h\nu}{E_u}$	η_{ult}, %
2833	$6p7s\,^3P_1^0 - 6p^2\,^3P_0$	0.21	0.57	18		Pump	
4057	$6p7s\,^3P_1^0 - 6p^2\,^3P_2$	0.15	1.05	9	6.95	0.70	44
3639	$6p7s\,^3P_1^0 - 6p^2\,^3P_1$	0.06	0.32	31	1.52	0.78	39
7229*	$6p7s\,^3P_1^0 - 6p^2\,^1D_2$	0.02	0.04	250	1.60	0.39	24
2170	$6p6d\,^3D_1^0 - 6p^2\,^3P_0$	0.39	1.84	5.5		Pump	
4062*	$6p6d\,^3D_1^0 - 6p^2\,^1D_2$	0.26	0.19	53	7.4	0.53	33
2613	$6p6d\,^3D_1^0 - 6p^2\,^3P_1$	0.02	1.10	9	0.33	0.83	42

does not apply to the transitions responsible for these two lines. On the other hand, we may assume that in order to obtain a considerable population inversion for these lines we need excitation pulses with steeper leading edges than those used experimentally. Since doubts have risen about the general population inversion mechanism, it would be very desirable to clear up the situation encountered in lead. Moreover, it would be interesting to achieve stimulated emission of new lead lines because such lines should have a high ultimate efficiency and should be located at shorter wavelengths than the 7229 Å line.

We were unable to find any published information on the electron-excitation cross sections of the atomic levels of lead. Therefore, a theoretical calculation of population inversion processes could hardly explain the situation. A calculation of the saturated power for lead was complicated by the occurrence of several interacting transitions. An estimate of the excitation cross sections of the resonance levels of lead, obtained using the simplified formula (30), gave $\sigma_{max}(^3P_1^0) = 3.7 \cdot 10^{-16}$ cm^2 and $\sigma_{max}(^3D_1^0) = 3.8 \cdot 10^{-16}$ cm^2, i.e., values close to those estimated for the resonance levels of thallium. From these estimates we expected population inversion for the 4062 Å line as well.

The populations of the ground and metastable levels of lead and the lead vapor pressures, obtained on the basis of [23], are given for various temperatures in Table 6.

It is clear from this table that the working lead vapor density is achieved at a temperature of 900-1000°C, i.e., higher than that of thallium but considerably lower than that of copper (1500°C). This is very important from the practical point of view because one can use fused quartz tubes for lead vapor.

Our experiments on lead were carried out using the same apparatus as that employed for thallium. In the first experiments we used a narrow tube of 1.3 mm diameter and 20 cm long containing the Pb^{208} lead isotope. Helium, neon, or krypton at a pressure of several Torr were used as the buffer gases. A resonator with two plane mirrors, located as close as possible to the tube, was employed. One mirror had a nontransmitting aluminized coating and the other was a substrate without any coating. The tube was excited with the pulse cable transformer described earlier.

Under these conditions we observed superradiance at 4057 Å beginning from 710-720°C. The intensity of this superradiant line increased with rising temperature. At about 750°C we observed laser emission and superradiance at 4062 Å, whereas at 760°C the same was noted for 3639 Å. Stimulated emission at 7229 Å appeared last at 820°C. All the lines exhibited a rise of the superradiance power with increasing temperature [58].

The wavelengths of the new laser emission lines were determined photographically using the STÉ-1 spectrograph. This determination gave the following values: 4057.79 ± 0.06, 4062.13 ± 0.06, and 3639.54 ± 0.06 Å. Within the limits of the stated error, all the lines agreed with the expected wavelengths of the transitions listed in Table 5.

Subsequent experiments demonstrated that all four stimulated radiation lines were easily observed under our conditions also for a natural mixture of lead isotopes. These four lines

TABLE 6

T, °K	T, °C	$N(^3P_0)$, cm^{-3}	$N(^3P_1)$, cm^{-3}	$N(^3P_2)$, cm^{-3}	p, Torr
900	627	$9.6 \cdot 10^{13}$	$1.1 \cdot 10^8$	$2.0 \cdot 10^6$	$9.0 \cdot 10^{-4}$
1000	727	$1.1 \cdot 10^{14}$	$4.1 \cdot 10^9$	$1.2 \cdot 10^8$	$1.1 \cdot 10^{-2}$
1100	827	$7.3 \cdot 10^{14}$	$7.9 \cdot 10^{10}$	$3.2 \cdot 10^9$	$8.4 \cdot 10^{-2}$
1200	927	$3.6 \cdot 10^{15}$	$9.2 \cdot 10^{11}$	$9.5 \cdot 10^{10}$	0.45
1300	1027	$1.4 \cdot 10^{16}$	$7.1 \cdot 10^{12}$	$5.1 \cdot 10^{11}$	1.87

behaved with temperature in the same way as the lines emitted by a single isotope but stimulated radiation appeared at temperatures about 100°C higher. This was clearly due to the fact that the gain for one of the isotopes should be approximately twice as high as for the other. Most subsequent measurements were carried out on a natural mixture of lead isotopes.

The characteristics of laser emission and superradiance due to transitions in lead, excited in tubes of 1.3 and 3 mm diameter, depended weakly on the nature and pressure of the buffer gas. However, the energies of the stimulated radiation pulses representing different lines varied in different ways with the voltage applied to the cable transformer. We determined the pulse energies because the time resolution of the apparatus was 10 nsec and the stimulated radiation pulses were considerably shorter.

Figure 36 gives the dependences of the energy of the stimulated radiation pulses on the voltage applied to the primary winding of the pulse cable transformer. It is clear from this figure that the nature of the dependence obtained for the 7229 Å line was quite different from that recorded for the other lines: at 850°C the laser emission at 7229 Å appeared at much lower voltages than for the other lines and the pulse energy reached its maximum value at voltages at which there was hardly any stimulated radiation due to the other lines. As the energy of the superradiance at 4057 Å increased with the voltage, the energy of the laser 7229 Å line decreased. At 900°C the superradiance at 4057 Å and the laser emission at 7229 Å appeared right from the beginning of the discharge, i.e., when the voltage applied to the transformer was 2-2.5 kV. Moreover, other lines also appeared at other voltages. At this temperature the energy of the pulses representing the laser emission at 7229 Å was practically independent of the voltage.

Somewhat different results were obtained in wider discharge tubes. Thus, when a tube of 11 mm internal diameter and an active zone 20 cm long was used, we found that a natural mixture of lead isotopes exhibited laser emission and superradiance only at two wavelengths: 4062 and 7229 Å. The buffer gas pressure influenced strongly the stimulated radiation characteristics. In particular, the addition of helium in amounts corresponding to 0-1 Torr resulted in predominance of the laser emission and superradiance at 4062 Å, whereas at higher pressures the 7229 Å line made the main contribution. Figure 37 gives the dependences of the energy of the stimulated radiation pulses on the helium pressure obtained at 900°C in a tube of 11 mm diameter and 20 cm long. It is clear from this figure that the nature of the pressure dependences obtained for these two lines was quite different.

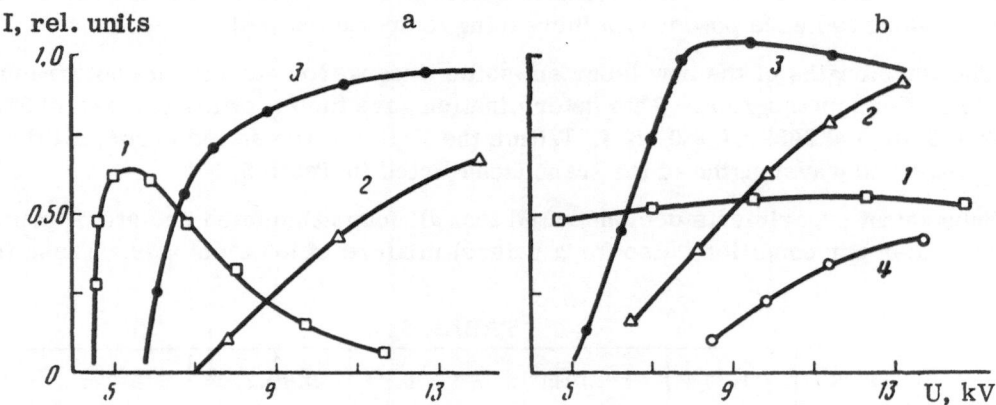

Fig. 36. Dependences of the energy of stimulated radiation pulses, corresponding to different lead lines, on the voltage applied to the cable transformer. The helium pressure was 2 Torr and the tube diameter was 2 mm. a) T = 850°C; b) T = 900°C; 1) 7229 Å; 2) 4057 Å; 3) 4062 Å; 4) 3639 Å.

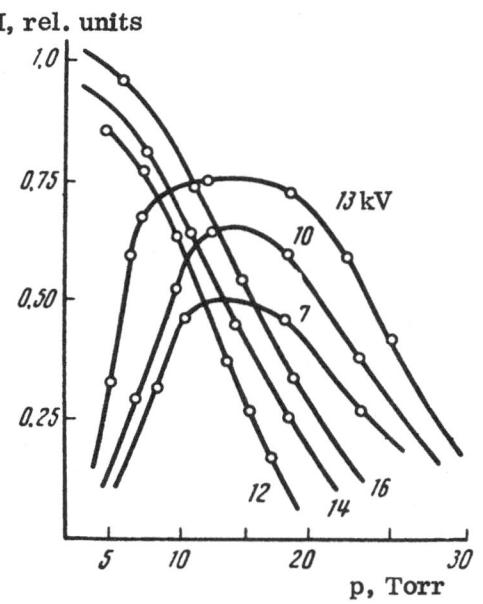

Fig. 37. Dependences of the energy of stimulated radiation pulses on the buffer gas (helium) pressure obtained for different voltages applied to the cable transformer (the voltages are given alongside the curves). The curves for 12, 14, and 16 kV apply to the 4062 Å line, whereas the curves for 7, 10, and 13 kV apply to the 7229 Å line.

Figure 38 gives the dependences of the energy of the stimulated radiation pulses on the temperature in the same tube at two helium pressures: a) 2.7 Torr, when the 4062 Å line predominated; b) 13 Torr, when the 7229 Å line was more important. It should be noted that at 900°C the laser emission regime at both wavelengths resulted in higher pulse energies than the superradiance regime, and the addition of a second mirror did not alter the pulse energy at temperatures above 900°C. It is clear from Fig. 38 that at the optimum (for each line) helium pressure the pulse energy of each line rose with temperature. However, when the helium pressure was 13 Torr, so that the 7229 Å line predominated, the energy of the 4062 Å pulses had a maximum at ~900°C.

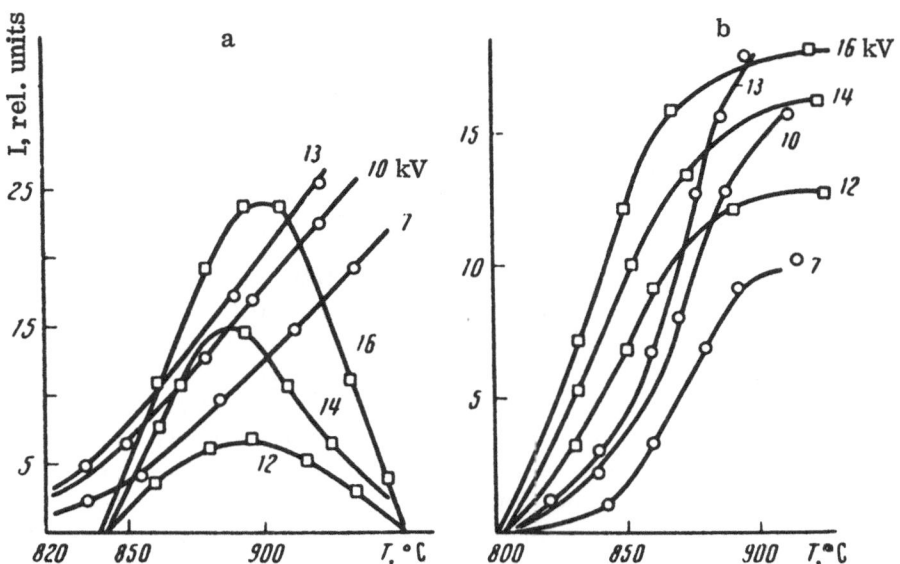

Fig. 38. Temperature dependences of the energy of stimulated radiation pulses generated in a tube of 11 mm diameter. Curves for 7, 10, and 13 kV apply to the 7229 Å line, whereas curves for 12, 14, and 16 kV apply to the 4062 Å line. a) Helium pressure 2.7 Torr; b) helium pressure 13 Torr.

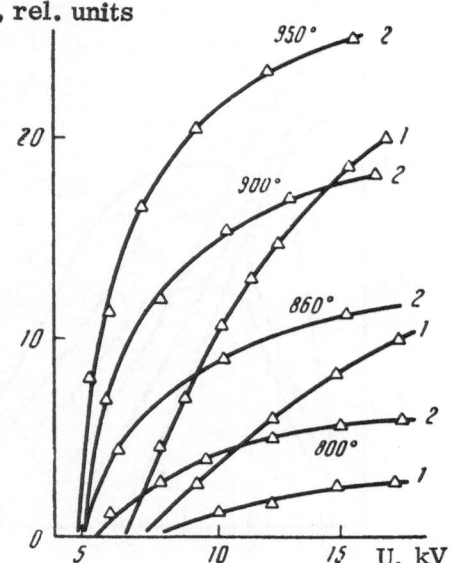

Fig. 39. Dependences of the laser emission energy at wavelengths 4062 Å (curves denoted by 1) and 7229 Å (curves denoted by 2) on the voltage applied to the cable transformer obtained at different temperatures from a tube of 11 mm diameter at a helium pressure of 6 Torr.

Figure 39 gives the dependences of the energy of the stimulated radiation pulses on the voltage applied to the transformer feeding the same tube; these dependences were obtained at various temperatures. It is clear that all the curves exhibited a tendency to saturation at higher voltages.

A comparison of the properties of stimulated radiation obtained as a result of excitation with pulses characterized by leading edges of different steepness could explain why laser emission of just one line was reported in [9, 54, 55]. With this in mind we carried out an investigation using a pulse generator without the cable transformer. In this case a capacitor was discharged via a thyratron through the discharge tube. The interval between the pulses was used to charge the capacitor and, therefore, the discharge tube was shunted by an inductance of several millihenries. When this type of power supply was used to excite a tube of internal diameter 1.3 mm, we observed laser emission of the 7229 Å line and the pulse energy was approximately the same as in the case when a cable transformer was used, but laser emission at the 4062 Å wavelength was very weak. No laser emission or superradiance was observed at 4057 and 3639 Å. In a tube of 11 mm diameter without the cable transformer we observed again only the 4062 and 7229 Å laser emission lines. Their energies and behavior was roughly the same as in the case when the tube was excited by the cable transformer.

These measurements were carried out with a time resolution of about 10 nsec. We found that the stimulated radiation pulses were also of about 10 nsec duration. Later we were able to carry out measurements with a time resolution of 10 nsec or shorter. In this case we used an FÉK-16 photocell in conjunction with an I2-7 oscillograph. The measurements were carried out on a discharge tube of 3 mm internal diameter and 16 cm long, filled with a natural mixture of the lead isotopes. The radiation emitted by the tube was not spectrally selected but was all concentrated by a lens on the photocathode of the coaxial photocell. The spectral composition of the radiation was determined with a UM-2 monochromator. Oscillograms of the stimulated radiation pulses obtained in this way are shown in Fig. 40. We found that at ~800°C the first to appear was the superradiance at 4057 Å. An oscillogram obtained in this case is shown in Fig. 40a. It was recorded at a buffer gas (neon) pressure of 5.5 Torr applying a voltage of 15 kV to the cable transformer. A mirror placed behind the tube at a minimum distance of 23 cm had no influence on the superradiance pulses. The duration of these pulses was ~1.2 nsec at midamplitude. Therefore, it was not surprising that the mirror had

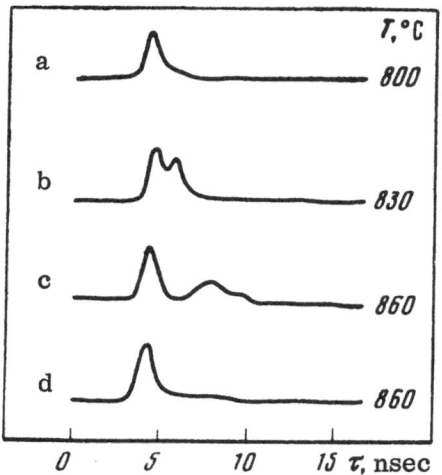

Fig. 40. Oscillograms of superradiance pulses emitted from a tube of 3 mm diameter and 16 cm active length at various temperatures.

no influence at this distance from the tube and for this pulse duration. It was not possible to bring the mirror closer to the tube because of the structural details of the apparatus. The influence of the mirror could be greater if the distance from the tube could be reduced to 5-10 cm.

When the temperature was increased further, we observed superradiance at 3639 Å (oscillogram in Fig. 40b). In this case again the mirror did not affect the observed superradiance. The same oscillogram shows a second peak, delayed by about 1.4 nsec. This peak was attributed to the 3639 Å line.

The oscillograms in Figs. 40c and 40d were obtained at a higher temperature of 860°C. In this case the superradiance peak at 3639 Å merged with the peak at 4057 Å but a new peak appeared at 4062 Å. This superradiance was influenced considerably by a mirror placed at a distance of 23 cm from the active zone. The oscillogram in Fig. 40c was obtained with a mirror and the one in Fig. 40d without a mirror. We found also that the 4062 Å superradiance pulse was considerably wider than the 4057 Å pulse. The duration of the former was 2.5 nsec at midamplitude and it appeared after a delay of 3.5 nsec relative to the 4057 Å pulse.

Figure 41 shows oscillograms of the 4057, 4062, and 7229 Å superradiance pulses generated in a tube of 5 mm diameter and 20 cm long at 900°C; the neon pressure was ~2 Torr and the voltage applied to the transformer was 7 kV. The spectral resolution was provided by a prism and this made it possible to record the red and blue radiations separately. It is clear from the oscillograms obtained that the 7229 Å laser emission pulse was delayed by 8-10 and 3 nsec relative to the 4057 Å and 4062 Å superradiance pulses, respectively. The beginning of

Fig. 41. Oscillograms of stimulated radiation generated in lead atoms: a) superradiance in laser emission at wavelengths 4057 and 4062 Å; b) spectrally unresolved radiation; c) laser emission at 7229 Å.

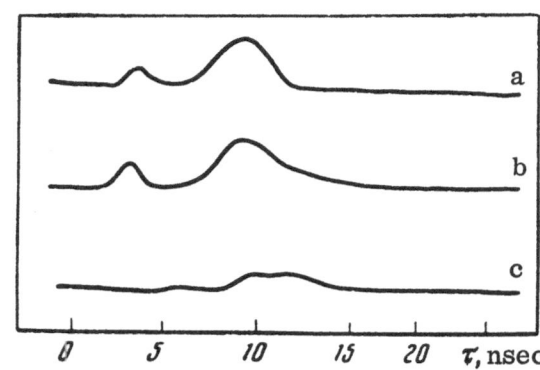

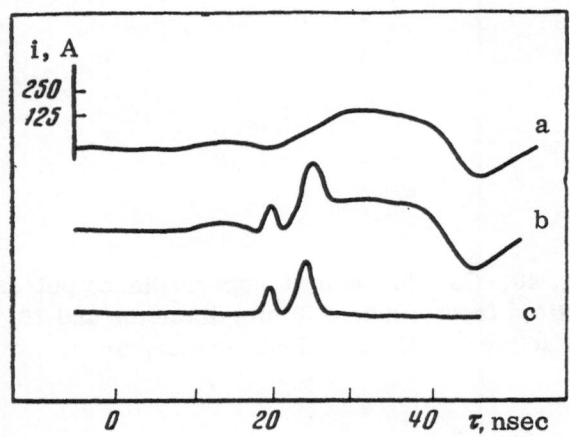

Fig. 42. Relative positions, on the time axis, of the current and lead stimulated radiation pulses: a) current pulse; b) current and stimulated radiation; c) superradiance at 4057 and 4062 Å wavelengths.

the laser emission at 7229 Å coincided approximately with the maximum of the laser emission at 4062 Å.

The relative positions of the current and stimulated radiation pulses (representing different lines) are shown in Fig. 42. The oscillograms in Fig. 42 were obtained under the same conditions as those in Fig. 41 except that the voltage applied to the transformer was 15 kV. Oscillograms were recorded by the same method as the corresponding oscillograms of thallium. It is clear from Fig. 42 that the 4057 Å superradiance appeared at the very beginning of the current pulse. Its appearance, relative to the current pulse, occurred at approximately the same point in time as the thallium superradiance. Superradiance and laser emission at 4062 Å were observed considerably later and were much longer, whereas the 7229 Å line appeared at the very end of the current pulse. The leading edge of the current pulse in these experiments was 12–15 nsec.

When the pulse cable transformer was not used, the leading edge of a pulse was considerably longer and it was 150 nsec for 15 kV. The current pulse amplitude was approximately the same for both excitation methods.

Figure 43 shows current and laser emission pulses obtained without the cable transformer. The voltage across the capacitor was 1.5 kV, the tube was of 5 mm diameter and 20 cm long, and the temperature was 900°C. In this case we observed only the laser emission of the 4062 and 7229 Å lines. The duration of this emission was about 5 nsec and it appeared at the very beginning of the current pulse.

Fig. 43. Relative positions, on the time axis, of the current and laser emission at wavelengths 4062 and 7229 Å obtained without the use of a cable transformer: a) current pulse; b) current and laser emission; c) laser emission.

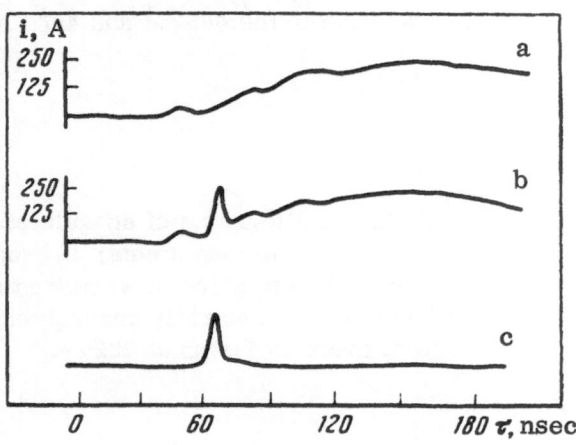

We used the same supply system (without the cable transformer) in measurements of the energy of the laser emission pulses at 7229 Å. When the capacitance was 2200 pF, voltage was 15 kV, repetition frequency was 5 Hz, neon pressure was 10 Torr, and discharge temperature was 900°C, the energy of the stimulated emission pulses was $4.2 \cdot 10^{-6}$ J. This corresponded to a peak power of ~1 kW for a pulse duration of 5 nsec (the measurements were carried out using a calibrated thermopile). Since the sensitivity of the coaxial photocell was not known accurately in this part of the spectrum, only rough estimates of the peak power could be obtained. These estimates gave approximately the same peak power as that just estimated. When the pulse repetition frequency was increased, the energy of the laser emission pulses fell somewhat and at 500 Hz it became $\sim1.6 \cdot 10^{-6}$ J. The average power of the 7229 Å line was then 0.8 mW. In the presence of the cable transformer the peak output power obtained under the same conditions was estimated to be about 2 kW.

The main result of our experiments on lead was the observation of stimulated emission of three new lines. All these lines were due to transitions from the resonance to metastable levels and were characterized by a high ultimate efficiency. The observation of population inversion for three new transitions indicated that the direct excitation of metastable levels of zinc was much less effective than the corresponding excitation of the resonance levels. Thus, lead confirmed once again the general hypothesis that the excitation cross sections of the metastable levels of atoms were small. Only one line was reported in [9, 54, 55] because the excitation pulses used in these investigations had insufficiently steep leading edges. A special feature of lead was that it exhibited many transitions from the resonance to metastable levels. These transitions had different probabilities and were located in different parts of the spectrum. Some of them shared the upper level and the others the lower level. Consequently, there was a mutual influence of the stimulated emission of the different lines. Investigations of laser emission and superradiance due to transitions in lead made it possible to study for the first time the nature of population inversion processes between different levels and to investigate the interaction between various lines. These results were important for the understanding of population inversion processes in other active media with large numbers of stimulated emission lines.

The following description of population inversion can be provided on the basis of the experimental results.

The metastable levels $6p^2\ ^3P_{1,2}$ are relatively low. At the working temperature their equilibrium populations reach considerable values and rise strongly with temperature. Population inversion occurs when a certain threshold pumping rate is exceeded so that during the lifetime of the upper level its population exceeds the population of the lower level. If the leading edge of the excitation pulses is sufficiently steep, this condition is satisfied for all the lines. Then, stimulated radiation in the form of superradiance appears first for the lines with the highest gain, for example, for the 4057 Å line. The duration of the superradiance pulses is governed by the processes discussed in Chap. I and under our experimental conditions it is controlled by collisions with electrons causing transitions between the active levels. The pulses are shortest for the 4057 Å line. When the population inversion of the 4057 Å line is exhausted, the conditions for the emission of stimulated lines due to transitions from the same upper level improve. Consequently, superradiance is observed at 3639 Å and then laser emission and superradiance at 7229 Å. This governs the relative times of appearance of the stimulated radiation pulses representing different lines.

If the leading edge of the excitation pulses is not sufficiently steep, the population is not inverted for the 4057 and 3639 Å lines. This does not mean that this inversion cannot be achieved for other transitions beginning from the same level. If, as a result of spontaneous radiation, the population of the metastable levels associated with the 4057 and 3639 Å lines rises sufficiently for the reabsorption of radiation emitted as a result of these transitions, the lifetime of the upper level $^3P_1^0$ is governed only by the low-probability decay involving the 7229-Å line. If the pumping is continued, population inversion is achieved for this line and all the pump energy is converted efficiently into stimulated radiation at 7229 Å. This is why stimulated emission of this line is easier at high temperatures and in wider tubes.

The 4062 Å line has to be considered separately. We can see from Table 5 that the product $\lambda\,^3A_{ik}$ for this line may be even greater than for the line at 4057 Å. Estimates of the excitation cross sections of the two resonance levels give approximately the same values (this point is discussed earlier). If allowance is made for the electron velocity distribution, it is found that the rate of excitation of the $^3D_1^0$ level should be somewhat less than that of the $^3P_1^0$ level. Moreover, the data for the 4062 Å line in Table 5 are the not very accurate results taken from [57], whereas the data for the 4057 Å line are reliable and taken from [56]. If we assume that the relative values of the transition probabilities given in [56] are closer to truth, we find that the transition probability producing the 4062 Å line is approximately one-third of the probability for the 4057 Å line. An indirect confirmation of this probability ratio was provided by the long delay and long duration of the superradiance pulses at 4062 Å. Since the stimulated emission conditions for this line were better in wider tubes, we concluded that reabsorption of the radiation from the upper level along all channels (with the exception of the active one) was important in this case.

Stimulated radiation due to transitions in lead atoms is also of considerable practical importance because it can be used to generate stimulated radiation in different parts of the spectrum and in the form of pulses of different duration. For example, very short pulses of about 1 nsec duration were observed at 4057 Å. On the other hand, at 7229 Å we could obtain pulses of much greater duration and a considerable average power (using the excitation system described above). Another important point was that different stimulated radiation lines were obtained in turn when the experimental conditions (temperature, pressure, buffer gas, discharge tube diameter, voltage, steepness of current pulses, and the resonator parameters) were varied.

The peak output powers at the red wavelength 7229 Å achieved in our experiments were not the maximum possible. The power could be increased further by the use of longer and wider discharge tubes. We did not estimate the efficiency of stimulated emission of the lead lines. However, it was clear that in practice a high efficiency would require the use of a system producing short and powerful excitation pulses.

§ 5. Superradiance in Mercury Vapor

It follows from the preceding sections that in the majority of atomic gas lasers one has to heat the active medium to high temperatures in order to obtain stimulated emission as a result of transitions from resonance to metastable levels. Clearly, it would be very desirable to achieve efficient stimulated emission from atoms at relatively low temperatures. Mercury is of special interest in this respect. The properties of mercury are well known because it is used in commercially produced electronic vacuum devices.

By analogy with the excitation processes in other atoms, we may expect pulse population inversion of some transitions in mercury under certain excitation conditions. In particular, by analogy with population inversion processes due to the 2p−1s transitions in neon (see Chap. IV), we may expect pulse population inversion of the $7^3S_1 - 6^3P_0$ (4047 Å) and $7^3S_1 - 6^3P_2$ (5461 Å) transitions as a result of multistage excitation of electrons via an intermediate resonance triplet level 6^3P_1. Stimulated emission at 5461 Å is of considerable interest because this line is used widely as a secondary wavelength standard and its precision is not much inferior to that of the primary standards. Stimulated emission at this wavelength would make it possible to simplify and refine the wavelength measurement process. We recall that mercury is particularly convenient for metrological measurements of wavelengths because its Doppler line width is relatively small and it has even isotopes free of hyperfine structure. Moreover, it should be possible to obtain stimulated emission as a result of other transitions in mercury. This was why we decided to investigate mercury vapor.

We used apparatus similar to that employed for thallium and lead. In the first experiments a mercury droplet was placed in a small side tube in the central part of the discharge. The discharge tube was made of quartz; its internal diameter was 1.3 mm and its active length was 20 cm. This tube was placed inside a heater. We employed an unsealed system with neon, helium, or argon as a buffer gas. The discharge in the tube was excited by the pulse cable transformer. The time characteristics of the stimulated and spontaneous radiation were investigated using an FÉU-36 photomultiplier and an S1-11 oscillograph with a time resolution of about 10 nsec.

We observed superradiance in the violet part of the spectrum [59] when the temperature in the heater was 60-80°C and the buffer gas (neon) pressure was in the range $6 \cdot 10^{-2}$-2.0 Torr. This superradiance appeared when the voltage applied to the transformer was 10 kV and the power increased with the voltage. We used one mirror which increased considerably the superradiance power. However, superradiance was not observed when helium or argon were used as the buffer gases. The working repetition frequency of the excitation pulses was 5-8 Hz; when this frequency was increased the superradiance became weaker.

We later found that in these first experiments the mercury was heated by the discharge so that its vapor pressure did not agree with the heater temperature. Therefore, we carried out subsequently experiments in which the discharge tube had a long side branch where the mercury was located. This side branch was heated to a definite temperature by a separate heater. The temperature in the working part was kept somewhat higher. Under these conditions we observed the superradiance in the temperature range 100-200°C, which corresponded to a mercury vapor pressure of 0.25-0.75 Torr. In this tube once again the superradiance was observed in the presence of helium in a narrow temperature range.

Figure 44a shows the structure of the spontaneous radiation spectrum at the wavelength which was observed also in superradiance; this structure was recorded employing a Fabry–Perot interferometer with a dispersion range of 2.5 cm^{-1}. We observed four components separated by intervals of about 0.5 cm^{-1}. The spectrum shown in Fig. 44b was obtained when superradiance just appeared. We can see that one of the components increased considerably in strength and became narrower. Figure 44c shows the spectrum obtained when the super-

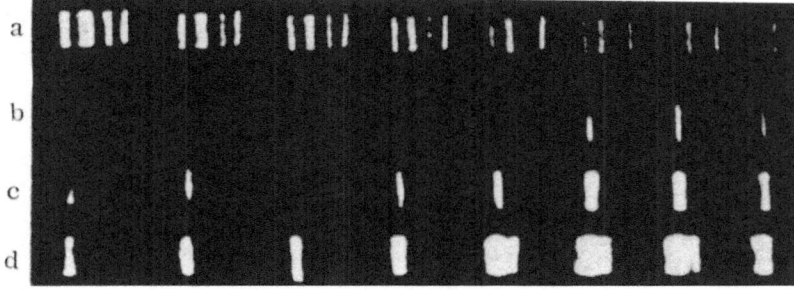

Fig. 44. Interferogram of 3984 Å spontaneous and stimulated radiation lines due to mercury ions: a) spontaneous radiation spectrum; b) low-power superradiance spectrum; c) maximum-power superradiance spectrum; d) maximum-power superradiance spectrum overexposed to show the weak component.

radiance power had its maximum value. The bright component in Fig. 44d was strongly over-exposed in order to show a weaker component. In some cases the superradiance was observed as three components with intervals of 0.5 cm^{-1}. The superradiance wavelength was measured by comparison with the wavelength of the iron line at 3983.96 Å using a Fabry—Perot inter-ferometer. In this case light emitted by a dc iron arc was directed to the same spot on the interferometer plates as the superradiance. The interference pattern was photographed at the exit from an STÉ-1 spectrograph. We used interferometers with dispersion ranges 1.0, 0.7, and 0.5 cm^{-1}. The measurements gave the wavelength 3983.99 ± 0.02 Å for the brighter com-ponent of the superradiance. Within the limits of the experimental error, this line coincided with the bright line of singly ionized mercury atoms, corresponding to the $6p^2P_{3/2}-6s^2\,^2D_{5/2}$ transition. A check made using tables in [44] indicated that, within the accuracy to which the level positions were known (about 1 cm^{-1}), this wavelength corresponded to three other tran-sitions:

$$\text{Hg II} \qquad B - 10^0_{1/2}\,(5d^96s6p),$$
$$\text{Hg II} \qquad A_{3/2} - 23^0_{1/2}\,(5d^96s6p),$$
$$\text{Hg III} \qquad 9^0_0\,(5d^96p) - 20^0_3\,(5d^86s6p).$$

The transition to Hg III was forbidden and, therefore, it was not very likely that super-radiance could be produced by this transition. The selection between the two transitions in Hg II was made using the data on the isotopic structure of the levels. The splitting for the even isotopes in the $6p^2P_{3/2}-6s^2\,^2D_{5/2}$ transition was 0.5 cm^{-1} [60, 61], whereas for the other two transitions it should be 0.3 cm^{-1} [62].

We concluded that the observed superradiance was due to the Hg II $6p^2P_{3/2}-6s^2\,^2D_{5/2}$ transition and the strongest component was due to the Hg^{202} isotope. The upper level of this transition was of the resonance type and the lower was coupled to the ground state of Hg II by the forbidden line 2814 Å. Population inversion in this transition could occur, on the assump-tion of pumping from the ground state of the ion, by the transitions from the resonance to meta-stable levels. The $ns^2\,^2D_{3/2}$ and $ns^2\,^2D_{5/2}$ levels associated with the excitation of an inner d electron were known to be exhibited not only by mercury ions but also by copper, gold, silver, as well as zinc and cadmium ions. In copper and gold atoms these levels were located below the resonance levels and, therefore, these atoms exhibited superradiance and laser emission due to transitions similar to those for mercury [7, 63], which provided an additional argument in support of the interpretation given above.

Under our experimental conditions there was no stimulated radiation at 5461 Å. The experimental data were insufficient to identify the cause of the absence of stimulated radiation at this wavelength. Such radiation could probably be obtained under different conditions. It would be interesting to continue investigations of mercury vapor using a wider range of tubes and experimental conditions. The use of a separated even isotope of mercury would increase considerably the probability of detection of stimulated radiation.

CHAPTER IV

PULSE LASER EMISSION AND SUPERRADIANCE DUE TO 2p—1s TRANSITIONS IN INERT GASES

The present chapter gives the results of an experimental investigation of stimulated radiation due to 2p—1s transitions (Paschen notation) in neon, argon, krypton, and xenon. All these inert gas atoms have similar level structures so that population inversion processes should be similar. As pointed out in the Introduction, the population inversion mechanism is not yet fully established. There are three points of view which are discussed in detail in the section below on neon. The main purpose of our investigation was to establish the population inversion mechanism. With this in mind we studied the properties of superradiance and laser emission due to transitions in neon. Stimulated radiation due to 2p–1s transitions in argon had not yet been reported. Therefore, we carried out a preliminary study in which such radiation was observed and investigated. The results obtained for superradiance emitted by neon and argon were used as the basis for conclusions on the population inversion mechanism.

§ 1. Superradiance and Laser Emission due to
Transitions in Neon

The special interest in neon arises from several factors. Laser emission from neon and its superradiance have been studied much more thoroughly than of other inert gases. Moreover, the characteristics of levels and transitions in neon, which are needed in the analysis of the processes under consideration, have been thoroughly investigated. Hence, it should be possible to identify more readily the population inversion mechanism in neon than in other inert gases. Finally, superradiance emitted by neon is of the greatest practical interest because the relevant wavelengths are located, in contrast to the other inert gases, in the visible part of the spectrum. It should also be pointed out that the highest peak power from a gas laser, amounting to 190 kW, has been obtained for the green line of neon [64, 65].

The neon level scheme is shown in Fig. 45. Out of 30 allowed 2p–1s transitions, three are active in laser emission and superradiance. We shall report observations of superradiance due to one additional transition. All these transitions are identified in Fig. 45 by arrows and are listed in Table 7. The duration of laser emission and superradiance pulses usually did not exceed 30 nsec. Both laser emission and superradiance were observed at the beginning of the current pulse. Superradiance of laser emission in the near infrared part of the spectrum were reported in [13, 14, 16, 17]. It was pointed out in [14] that laser emission of

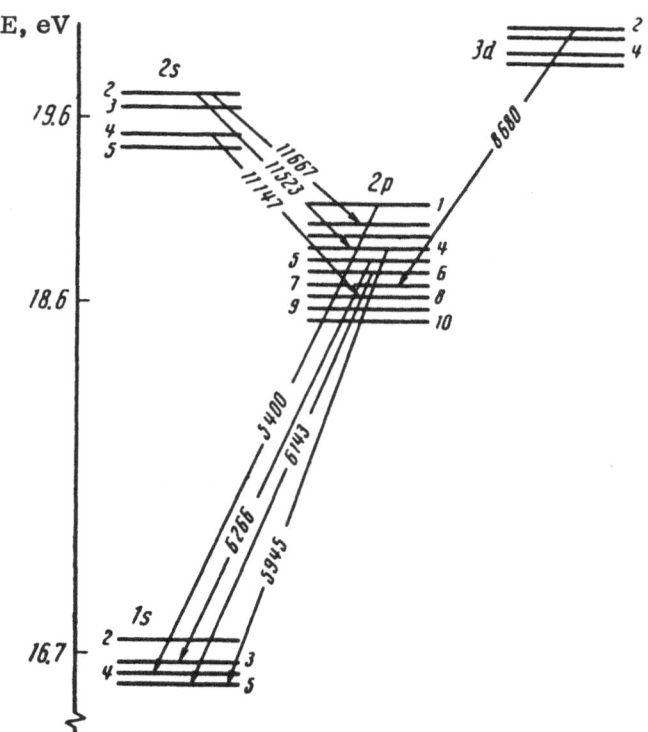

Fig. 45. Energy level scheme of neon (Paschen notation). The arrows identify the transitions which give rise to superradiance and laser emission.

TABLE 7

Gas	λ, Å	Transition	Reference	Gas	λ, Å	Transition	Reference
				Xe	9045	$2p_9$—$1s_5$	17
					9799	$2p_{10}$—$1s_5$	17
Ne	5400	$2p_1$—$1s_4$	14, 15, 16, 64, 65, 66		8410	$2p_8$—$1s_5$	17
	5945	$2p_4$—$1s_5$	13, 14, 16, 66	Ar	7634.8	$2p_6$—$1s_5$	66
	6143	$2p_6$—$1s_5$	13, 14, 16, 66		7723.6	$2p_7$—$1s_5$	66
	6266	$2p_5$—$1s_3$	Our results		7948.0	$2p_4$—$1s_3$	66
Kr	8104	$2p_8$—$1s_5$	17		6965.8	$2p_2$—$1s_5$	67
					7067.3	$2p_3$—$1s_5$	67

infrared lines began somewhat earlier than at the visible wavelengths. The transitions responsible for the infrared superradiance were not known.

Three different points of view have been put forward on the population inversion mechanism due to the 2p—1s transitions in neon. It was suggested in [13, 14] that inversion of the 2p—1s transitions was due to radiative cascades mainly from the 2s levels and it was estimated [13] that the observed peak power required also stimulated emission due to the 2s—2p transitions.

A mechanism of multistage excitation of the 2p levels by electrons via an intermediate resonance $1s_2$ was put forward in [16]. A similar point of view was recently advanced in [68] on the basis of an analysis of the lifetimes and excitation cross sections of the neon levels. Finally, it was suggested in [15, 64] that the upper level was excited from the ground state of the neon atom by direct electron impact. The predominance of any one of these three mechanisms should depend primarily on the relative excitation cross sections of the neon levels by electrons.

The cross sections for the direct excitation, from the ground state, of the levels of interest to us are listed in Table 8. This table gives the designations of the neon levels in terms of the LS and jl couplings, and also in accordance with the Paschen notation, as well as the

TABLE 8

Level				Effective cross section at maximum in units of 10^{19} cm^2			
LS	jl	Paschen notation	E, eV	[69]	[70]	[72]	[73]
$3s\ {}^1P_1^0$	$3s'\ [1/2]_1^0$	$1s_2$	16.844				28.0
$3s\ {}^3P_0^0$	$3s'\ [1/2]_0^0$	$1s_3$	16.723				2.04
$3s\ {}^3P_1^0$	$3s\ [3/2]_1^0$	$1s_4$	16.667				3.4
$3s\ {}^3P_2^0$	$3s\ [3/2]_2^0$	$1s_5$	16.616				2.6
$3p\ {}^1S_0$	$3p'\ [1/2]_0$	$2p_1$	18.961	6.0	38.9		8.58
$3p\ {}^3P_1$	$3p'\ [1/2]_1$	$2p_2$	18.734	1,4	5.17		0.03
$3p\ {}^3P_0$	$3p\ [1/2]_0$	$2p_3$	18.719	1.1	3.4		0.57
$3p\ {}^3P_2$	$3p'\ [3/2]_2$	$2p_4$	18.699	3.2	11 (6.5)		0.18
$3p\ {}^1P_1$	$3p'\ [3/2]_1$	$2p_5$	18.689	2.0	5.78		0.005
$3p\ {}^1D_2$	$3p\ [3/2]_2$	$2p_6$	18.644	2.9	16 (12.2)		1.8
$3p\ {}^3D_1$	$3p\ [3/2]_1$	$2p_7$	18.608	2.3	6.33		0.49
$3p\ {}^3D_2$	$3p\ [5/2]_2$	$2p_8$	18.586	3.6	14 (11.2)		0.54
$3p\ {}^3D_3$	$3p\ [5/2]_3$	$2p_9$	18.563	2.5	12.4 (8)		1.12
$3p\ {}^3S_1$	$3p\ [1/2]_1$	$2p_{10}$	18.317	2.0	11 (5)		2.3
$4s\ {}^1P_1^0$	$4s'\ [1/2]_1^0$	$2s_2$	19.69			7.1	1.66
$4s\ {}^3P_0^0$	$4s'\ [1/2]_0^0$	$2s_3$	19.67				0.16
$4s\ {}^3P_1^0$	$4s\ [3/2]_1^0$	$2s_4$	19.59				2.21
$4s\ {}^3P_2^0$	$4s\ [3/2]_2^0$	$2s_5$	19.57				0.30

level energies in electron-volts and excitation cross sections at the maxima (in units of 10^{-19} cm^2). Table 8 gives only the results of recent investigations with the fullest data. In particular, comprehensive measurements of the level parameters of interest to us were reported in [69, 70]. It should be noted that experimental determination of the direct electron excitation cross sections of the 2p levels is difficult because the contribution of the cascade transitions can be identified only by carrying out measurements on a large number of lines in the infrared part of the spectrum. In view of these difficulties the cascade transitions were not excluded at all in [69] and were allowed for only partly in [70]. However, direct electron-beam experiments [68, 71] demonstrated conclusively that cascade transitions, primarily from the 2s level, played a very important role in the population of the 2p levels. This was indicated also by the results of an investigation of the population inversion processes involving the 2s −2p transitions in neon [74-76], particularly the occurrence of continuous and pulse laser emission from pure neon at low currents and pressures. Even the results reported in [70] suggested that the main contribution to the measured cross sections was made by the cascade transitions. In fact, the energy dependences of the cross sections obtained in [69, 70] had wide maxima and gently sloping wings typical of the allowed transitions, whereas the dependences representing the direct excitation of the 2p levels should have sharper maxima. The most important transitions for the majority of the 2p levels were those from the 2s levels and these were not allowed for. Only in the sole case of the $2p_{10}$ level was an allowance (and then only partial) made for the cascades from the 2s level and this indicated that the measured cross section for the $2p_{10}$ level should be reduced by a factor of about 3. It should also be noted that the available information on the excitation cross sections of the 2s levels (Table 8) and transition probabilities [68] indicated that a contribution of the cascade transitions to the excitation cross sections of the 2p levels was comparable with the measured cross sections reported in [69, 70]. Moreover, a comparable contribution could also be made by cascades from the 3d levels.

Thus, the available experimental information on the direct-excitation cross sections of the 2p levels of neon should be regarded as giving the upper limits of the cross sections. The excitation cross sections of the 2s levels were determined experimentally only in the $2s_2$ case [72] and, to the authors' knowledge, the excitation cross sections of the 1s levels had not been measured.

The electron excitation cross sections of the neon levels have been calculated theoretically in several papers. The fullest calculations of the oscillator strengths and electron excitation cross sections of many levels of interest to us were reported in [73]. These calculations were carried out in the intermediate coupling approximation. The results obtained are also included in Table 8.

Multistage electron-excitation cross sections can be estimated if information is available on the excitation of the 2p levels from the 1s levels. Experimentally measured excitation cross sections of some of the transitions to the 2p levels directly from all the 1s levels were given in [69]. The same cross sections were measured in [77] by a somewhat different method and an attempt was also made to obtain the excitation cross sections of the 2p levels separately from each 1s level on the assumption of proportionality of the cross sections to the corresponding oscillator strengths. Since the calculations reported in [73] and general considerations suggested that the $1s_2$ resonance level was excited by electrons much more effectively than the other 1s levels, we concentrated our attention on the excitation cross sections of the 2p levels from the $1s_2$ level. Table 9 gives the cross sections obtained in [77] and the values deduced on the basis of the experimental data [69] by calculations analogous to those described in [77]. It is clear from Table 9 that the cross sections differed on the average by a factor of 2, which should be regarded as satisfactory in view of the experimental difficulties, inaccuracies in the calculations [69], and the fact that under the experimental conditions used in the studies reported in [69, 77] the process of filling of the majority of the 2p levels was dominated not by the $1s_2$ level but by the $1s_5$ level.

TABLE 9

Level	σ_{max}, 10^{-16} cm^2			Level	σ_{max}, 10^{-16} cm^2		
	[77]	[69]	[73]		[77]	[69]	[73]
$2p_1$	69	154	6.4	$2p_6$	64	33	15
$2p_2$	17	15	8.8	$2p_7$	0	0	4.7
$2p_3$	0	0	0.88	$2p_8$	14	6	6.1
$2p_4$	23	48	8.8	$2p_9$	0	0	19
$2p_5$	21	13	16.1	$2p_{10}$	0	0	8.8

The experimentally determined excitation cross sections of the 2p levels from the $1s_2$ level were very large for the allowed transitions: They amounted to 10^{-15}-10^{-14} cm^2. Table 9 gives the cross sections calculated in [73]. These were somewhat smaller and agreed much less satisfactorily with the experimental results than the experimental results themselves derived from different sources.

Comparing all the available data, we concluded that the excitation cross sections of the 2p levels from the ground state were similar and their order of magnitude was 10^{-19} cm^2. The excitation cross sections of the $1s_3$, $1s_4$, and $1s_5$ levels found by calculation were of the same order of magnitude and the cross section of the $1s_2$ level was considerably greater. The excitation cross sections of the $2s_2$ and $2s_4$ levels were slightly greater than those of the 2p levels. On the basis of these estimates we expected pulse population inversion at the leading edge of the excitation pulses due to the 2s−2p transitions, which was confirmed experimentally. Population inversion due to direct excitation of the 2p−1s transitions was hardly possible. Moreover, our estimates indicated that it should be possible to achieve pulse population inversion of the 2p−1s transitions due to stimulated cascades from the 2s levels and due to multistage excitation of electrons via the $1s_2$ intermediate level. It was difficult to choose between the cascade and multistage mechanisms.

Unfortunately, the above conclusions were not very reliable. Clearly, the available experimental data on the cross sections were incomplete and did not allow sufficiently for the influence of the cascades, and the results obtained by different workers disagreed quite considerably. The degree of reliability of the approximations used in the theoretical calculations had not yet been analyzed. The agreement between the calculated and experimental values could not be regarded as fully satisfactory. In this situation we could hardly solve the problem by calculating population inversion. Therefore, we concentrated our attention on the experimental study.

The apparatus used in our experiments was similar to that described in Chap. II and it is shown in Fig. 9. Superradiance and laser emission from neon were studied using four tubes. Three of them were 20 cm long with internal diameters of 1.3, 3, and 5 mm; the fourth had an active length of 120 cm and a diameter of 6.5 mm. Only cold electrodes were used. We noted that superradiance increased considerably when the cathode was a hollow cylinder. Therefore, we used a hollow Kovar cylinder as the cathode and a massive tungsten electrode, taken from an IFP-2000 lamp, as the anode. The discharge in the tubes was excited by the cable pulse transformer described earlier.

The radiation emitted from the discharge tubes was investigated under two sets of conditions: 1) laser emission, i.e., in the presence of a resonator composed of two mirrors; 2) superradiance, i.e., with only one mirror behind the tube or without a mirror at all. In most of the investigated cases the intensity of the stimulated radiation lines in the presence of a resonator was less than under superradiance conditions. It should be pointed out that the presence of one mirror had a significant influence on the superradiance intensity only for some lines and only under certain conditions. The results reported below were obtained using just one mirror behind the tube, unless stated to the contrary.

The results obtained depended strongly on the discharge tube diameter. In a narrow tube, of 1.3 mm diameter and 20 cm long, visible superradiance was observed at three lines:

Wavelength λ, Å	6143	5944.8	6266.5±0.1
Transition	$2p_6-1s_5$	$2p_4-1s_5$	$2p_5-1s_3$
	(strong)	(weak)	(weak)

The wavelength of the line observed first, 6266 Å, was measured photographically using a DFS-13 spectrograph with a dispersion of 2 Å/mm. Within the limits of the error given above, this line could be attributed only to the $2p_5-1s_3$ transition. The wavelengths of the other lines were determined photoelectrically using a DFS-12 spectrograph. No superradiance or laser emission was observed at the wavelength of the 5400-Å green line of neon in the narrow tube under any of the conditions used in our experiments. In addition to the visible superradiance, we found that the same tube emitted also infrared superradiance at the following lines:

Wavelength λ, Å	11523	11767	11143	8681
Transition · · · · · · · · ·	$2s_2-2p_4$	$2s_2-2p_2$	$2s_4-2p_8$	$3d_2-2p_7$
	(strong)	(strong)	(strong)	(weak)

Figure 46a shows the dependences of the energy of superradiance pulses on the neon pressure in the tube of 1.3 mm diameter with an active length of 20 cm. The voltage applied to the primary winding of the transformer was 7 kV and the repetition frequency was 100 Hz. Curve 1 is plotted for the 6143 Å line. It represents the behavior of the superradiance lines in the visible part of the spectrum: the 5945 and 6266 Å lines behaved analogously except that the pulse energy of these lines was considerably less. Curve 2 applies to the 8681 Å line and curve 3 to the 11,523 Å line; the other superradiance lines due to the 2s−2p transitions exhibited similar dependences. It is clear from Fig. 46a that the behavior of the superradiance lines belonging to different transition groups was quite different. The optimal pressure for the lines in the 2p−1s group was 1.3 Torr, whereas the optimal pressures for the 3d−2p and 2s−2p lines were 2.4 and 5 Torr, respectively.

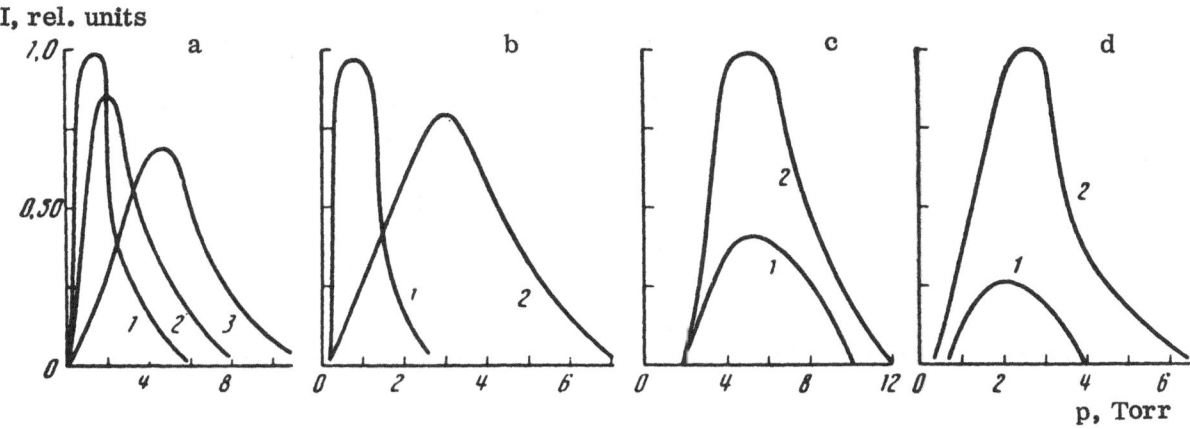

Fig. 46. Dependences of the energy of superradiance and laser emission pulses on the neon pressure: a1) 6143 Å line; a2) 8681 Å; a3) 11,523 Å; b1) 6143 Å; b2) 11,523 Å; c1) 5400 Å superradiance line; c2) 5400 Å laser emission line; d1) 11,523 Å laser emission line; d2) 11,523 Å superradiance line.

The use of a mirror in conjunction with the narrow tube increased the superradiance pulse energy at 6143 and 5945 Å. Here, as in the experiments with thallium, a mirror influenced the superradiance pulse energy only for certain distances between the mirror and the active part of the tube (detailed information of this point will be given later). The dependences mentioned above were obtained when this distance was 30 cm. The 5945 Å line was affected only when the mirror was placed almost next to the tube end. However, the superradiance pulse energy increased only at certain neon pressures. In the case of the 6143 Å line the increase was observed in the pressure range 0.1–1.5 Torr, whereas in the case of the 5945 Å line it was observed from 1 to 2 Torr. In the presence of a mirror the optimal pressure for the 6143 Å line decreased to 0.6 Torr. When a mirror influenced the superradiance energy, it also reduced strongly the size of the superradiant spot. Thus, under optimal conditions a mirror increased the energy of the superradiance pulses at the 6143 Å wavelength by a factor of about 25 near the center of the spot, whereas the total energy increased only by a factor of 4.

In the wider tube of 5 mm diameter and 20 cm long we observed the same superradiance lines as in the narrow tube, with the exception of the 6266 Å line. Moreover, we found superradiance and laser emission at the 5400 Å $(2p_1-1s_4)$ wavelength. In contrast to the tube of 1.3 mm diameter, we now found that a mirror behind the tube increased the energy of stimulated radiation pulses at all wavelengths and throughout the range of their emission, apart from the 5945 Å line. This line was not affected at all by the presence of a mirror. Therefore, Fig. 46b shows the dependence of the energy of the superradiance pulses on the neon pressure in the presence of a mirror located at a distance of 30 cm from the active part of the tube. The voltage applied to the pulse cable transformer was 7 kV and the pulse repetition frequency was 100 Hz. Curve 1 represents the behavior of the visible superradiance at wavelengths 6143 and 5945 Å. The optimal pressure for these lines was 0.6 Torr. Curve 2 represents infrared lines due to the 2s−2p transitions. In this case the optimal pressure was 3 Torr. Laser emission at the 5400 Å wavelength was obtained from this tube in the pressure range 2–10 Torr with a broad maximum near 5 Torr.

More powerful laser emission and superradiance were obtained at 5400 Å from a tube 120 cm long of 6.5 mm diameter. Figure 46c shows the neon pressure dependence of the energy of stimulated radiation pulses emitted at this wavelength. The voltage applied to the transformer was 12 kV and the repetition frequency was 15 Hz. Curve 1 was obtained under superradiance conditions and curve 2 was obtained for laser emission. The optimal pressure was about 5 Torr. We used spherical mirrors of 2 m radius of curvature and aluminized coatings. One mirror was nontransmitting and the other had a transmission of 10–15%. It is clear from Fig. 46c that a high power was obtained under laser emission conditions. Superradiance or laser emission due to other visible lines were not observed for this type, which was 120 cm long and of 6.5 mm diameter. In the infrared range we observed the same superradiance lines as those emitted by the other tubes. The dependences of the energy of infrared pulses on the neon pressure are plotted in Fig. 46d. Curve 1 represents laser emission and curve 2 represents superradiance. The conditions were the same as for the laser emission of the 5400 Å line.

The behavior of the superradiance lines belonging to different groups of transitions also varied with the voltage applied to the cable transformer or with the discharge current. Figure 47a shows the dependence of the amplitude of the current pulses on the voltage applied to the primary winding of the transformer (the discharge took place in a 5 mm diameter tube, 20 cm long and with $p_{Ne} = 0.7$ Torr), whereas Fig. 47b gives the dependence of the energy of superradiance pulses emitted from a tube 1.3 mm in diameter and 20 cm long on the voltage applied to the transformer. The neon pressure was 2.3 Torr. Figures 47c and 47d give analogous dependences for a tube 5 mm in diameter and 20 cm long. The neon pressures were 0.7

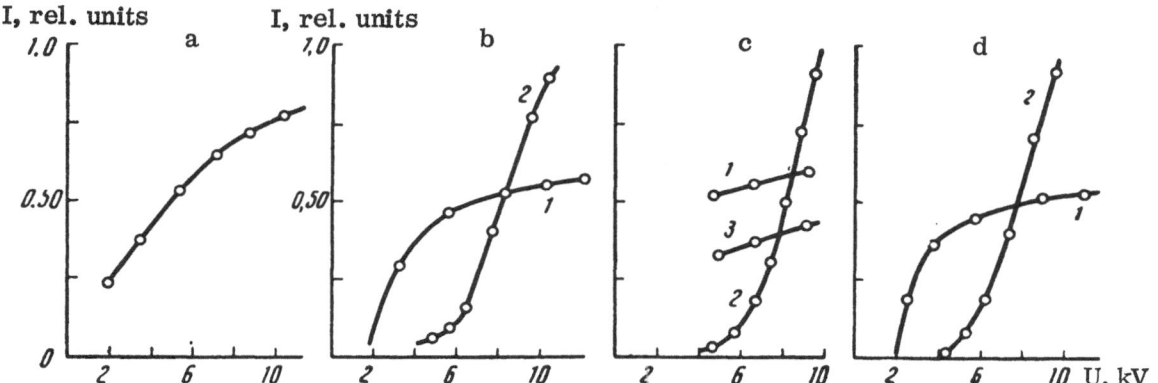

Fig. 47. Dependences of the current pulse amplitude (a) and superradiance pulse energy (b-d) on the voltage applied to the pulse cable transformer: 1) 11,523 Å line; 2) 6143 Å; 3) 11,143 Å.

Torr (Fig. 47c) and 0.2 Torr (Fig. 47d). All the curves in Fig. 47 were obtained using one mirror placed at a distance of 30 cm from the active part of the discharge tube. We found that infrared superradiance began at much lower voltages and currents than the superradiance due to the 2p−1s transitions. A steep rise of the energy of the superradiance pulses due to the 2p−1s transitions started only when the lines in the 2s−2p group reached saturation. When the voltage on the primary winding of the transformer was increased to 10-15 kV, we observed also saturation of the 2p−1s transitions.

An oscillographic study of superradiance pulses, carried out with the aid of FÉU-36 and FÉU-28 photomultipliers, demonstrated that the duration of all the visible and infrared pulses was 10-15 nsec and it did not vary greatly with the experimental conditions. Our apparatus failed to reveal any time delay of the radiation due to the 2s−2p and 2p−1s transitions. Since the resolution of the apparatus was estimated to be 10-15 nsec, there could be some delay not exceeding 10 nsec.

Figure 48 shows oscillograms of superradiance pulses obtained at the 6143 Å wavelength with a time resolution of 1 nsec in a tube 3 mm in diameter and with an active length of 20 cm; these oscillograms were recorded at different neon pressures. A mirror was placed at a distance of 10 cm from the active part of the discharge tube and the voltage applied to the transformer was 15 kV. The superradiance pulses were complex and their shape depended on

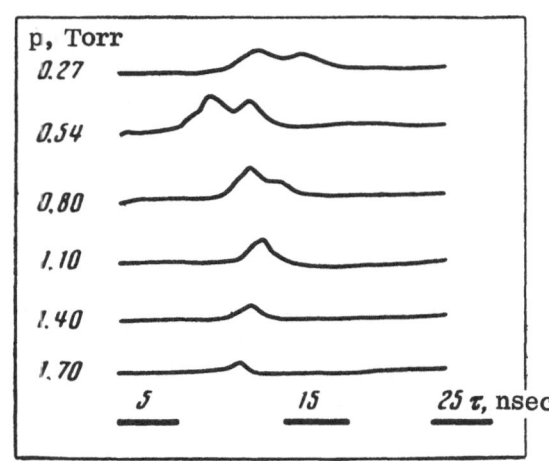

Fig. 48. Influence of the neon pressure on oscillograms of the 6143 Å superradiance pulses generated in a tube 3 mm in diameter and 20 cm long.

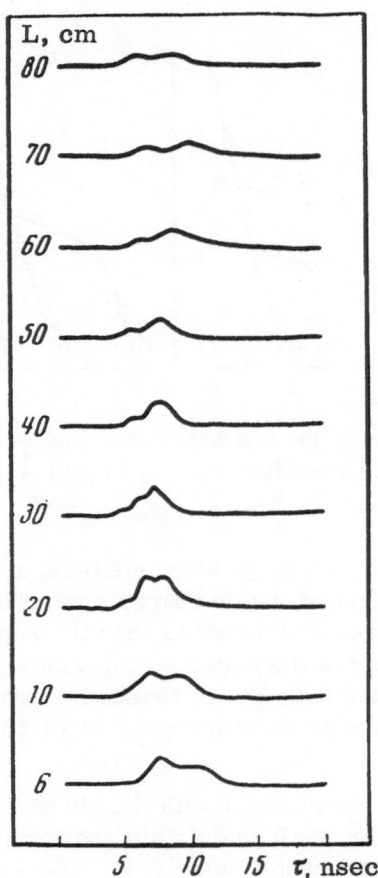

Fig. 49. Influence of the distance between a mirror and the active medium on the oscillograms of the 6143 Å superradiance pulses.

the neon pressure. When this pressure was increased, the shape became simpler and the pulses became shorter. For example, at a pressure of 0.27 Torr, the pulse duration at midamplitude was 5 nsec, whereas at 1 Torr it was 1.5 nsec. When the pressure was increased still further, the pulses became much shorter and their amplitude decreased considerably. Therefore, these pulses could not be detected directly at high pressures.

Figure 49 gives oscillograms of superradiance pulses obtained for different distances between a mirror and the active part of the tube. The neon pressure was 0.27 Torr and the other conditions were the same as those for the curves in Fig. 48. Clearly, the pulse amplitude and shape depended strongly on the distance of the mirror from the tube. The influence of the mirror was limited to the appearance of an additional power peak, which shifted along the superradiance pulse when the mirror position was varied. This shift corresponded, to within the experimental error, to the time needed for the propagation of light from the active part of the tube to the mirror and back again. When the propagation time became greater than the duration of the superradiance pulse, the additional peak disappeared and the mirror ceased to influence the superradiance pulses.

Figure 50 gives the neon-pressure dependence of the distance from the active part of the tube to the mirror at which the mirror ceased to influence the superradiance power. This figure includes also the dependence of the duration of the superradiance pulses (measured at the base) on the neon pressure. Clearly, when the scales were suitably selected, these dependences were very close to one another. Thus, the dependence of the superradiance power on the distance from the mirror could be used to estimate quite accurately the duration of nanosecond superradiance pulses. It should be pointed out that the influence of the mirror on

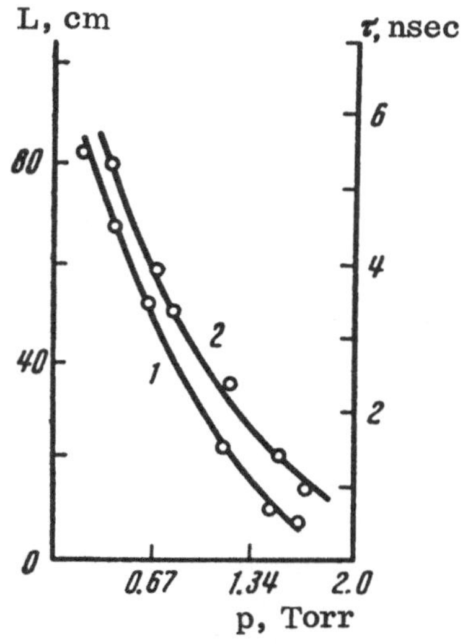

Fig. 50. Dependences, on the neon pressure, of the maximum distance between a mirror and the active part of the tube at which the mirror ceases to influence superradiance pulses (curve 1) and of the duration of oscillographically measured superradiance pulses (curve 2).

the superradiance power was detected using an instrument with a relatively poor time resolution. Consequently, one should be able to use slow-response detectors in measurements of the duration of nanosecond superradiance pulses.

Figure 51 shows the positions, on the time axis, of the 6143 Å superradiance pulses, relative to the current pulses applied to a tube whose active length was 20 cm and whose diameter was 5 mm. The experimental conditions were the same as those for the curves in Fig. 48. For comparison, we included in Fig. 51a the corresponding oscillogram of the superradiance of thallium vapor. Clearly, the neon superradiance appeared at the beginning of the current pulse but it was delayed slightly relative to the thallium superradiance.

A coaxial photocell was used in measurements of the peak superradiance power at the wavelength of the 6143 Å neon line. The maximum power recorded for a tube which was 3 mm in diameter and which had an active length of 20 cm was 60 W when the neon pressure was 0.6 Torr and the voltage applied to the transformer was 15 kV; this corresponded to a specific peak power of 45 W/cm^3. In this case, the duration of the pulses was 1 nsec.

The experimental results reported above provided a basis for identifying the population inversion mechanism in the case of the 2p−1s transitions in neon. The cascade population of the upper active levels suggested in [13, 14] is not in agreement with the experimental results. As pointed out earlier, the cascade population of the 2p levels originates largely from the 2s

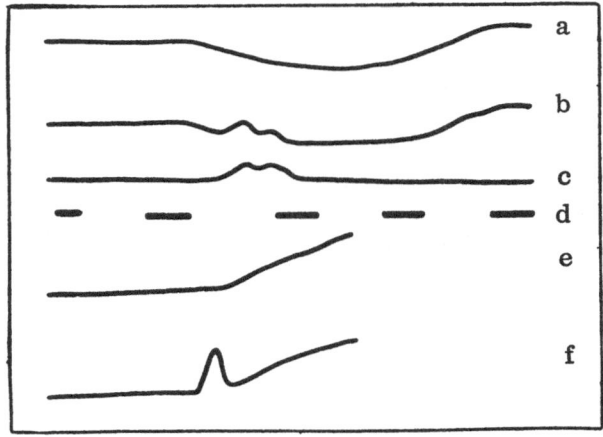

Fig. 51. Relative positions, on the time axis, of the current and 6143 Å superradiance pulses: a) discharge current; b) current and superradiance in neon; c) superradiance in neon; d) 10-nsec timing marks; e) discharge current in thallium vapor; f) current and 5350 Å thallium superradiance pulses.

levels, and the contribution of the 3d levels is less. The $2s_1$ and $2s_4$ levels, coupled optically to the ground state of the neon atom, are the 2s levels populated most effectively by direct electron impact. Under our experimental conditions, these levels generate superradiance terminating at some 2p levels; moreover, superradiance from the $3d_2$ level is observed.

The probabilities of the 2s–2p spontaneous transitions are of the order of 10^7 sec^{-1}. During stimulated radiation pulses resulting from the 2p–1s transitions (these pulses last about 5 nsec), the pumping by spontaneous cascades is less important than that due to stimulated cascades. This means that, under our conditions, it is sufficient to consider only stimulated cascades. Clearly, the rate of population is highest for those 2p levels which are the final levels of the superradiant transitions from higher levels. It is clear from Fig. 45 and Table 7 that the infrared superradiance from the 2s and 3d levels terminates at the $2p_2$, $2p_4$, $2p_7$, and $2p_8$ levels, whereas the superradiance in the visible part of the spectrum begins from the $2p_1$, $2p_4$, $2p_5$, and $2p_6$ levels. Thus, three out of four superradiance lines due to the 2p–1s transitions cannot be explained at all by radiative cascades. The 5945 Å ($2p_4$–$1s_5$) line needs special consideration. Inversion of the transition responsible for this line may result from cascade pumping due to the 11,523 Å ($2s_2$–$2p_4$) superradiance. However, the absence of a definite correlation between the behavior of these two lines shows that the cascade mechanism is not the main cause of population inversion even in the case of the 5945 Å line. In particular, the voltage dependences of the infrared and visible superradiance power are quite different (Fig. 47) and show no correlation. In the cascade laser emission, a power maximum for one of the cascade lines gives rise to a maximum of the other line at the same pressures [78, 79]. Moreover, the dependences of the superradiance power on the voltage applied to the tube are different for the infrared and visible lines (Fig. 47). Furthermore, the influence of a mirror is different for the infrared and visible lines. Thus, all the experimental results indicate that the population of the 2p levels by radiative cascades from above makes no significant contribution to population inversion of the 2p–1s transitions in neon.

We now have to select between the direct and multistage population of the upper active levels by electron impact. Under our experimental conditions in the presence of a strong superradiance due to the 2s–2p transitions, i.e., close to saturation, we may assume that the rate of pumping of those 2p levels which give rise to strong superradiance lines is practically the same as the rate of pumping of the $2s_2$ level. However, as shown above, the cascade superradiance does not occur under our experimental conditions. Thus, the rate of pumping by direct electron impact with an excitation cross section of about $7 \cdot 10^{-19}$ cm^2 (this applies to the $2s_2$ level) is insufficient for the population inversion of the 2p–1s transitions. The above values of the excitation cross sections of the neon levels from the ground state of the atom (we recall that the experimental values are overestimated because they ignore cascades) suggest that the excitation cross section of the $2s_2$ level is greater than the cross sections of the 2p levels. This is supported convincingly also by the continuous and pulse laser emission due to the 2s–2p transitions.

Consequently, if the pumping by a stimulated cascade does not give rise to a population inversion of the 2p–1s transitions, the direct electron excitation of the 2p levels is even less likely to produce this population inversion. Moreover, if we assume that the 2p levels are populated by direct electron impact, we find that the ratio of the rates of population of the 2s and 2p levels should be constant when the electron density is varied because the $2s_2$ and $2s_4$ levels are, according to all the available data, populated by direct electron impact. In this case, the infrared superradiance lines should behave like the visible lines when the discharge current is varied. However, it is often found experimentally that there is a considerable difference between the dependences of the parameters of these lines on the discharge current (Fig. 47). Moreover, the superradiance due to the 2p–1s transitions appears at higher discharge currents than the infrared superradiance and it rises more radiply with the current.

The probabilities of the 2p−1s transitions are an order of magnitude greater than those of the 2s−2p transitions. This means that, for a comparable pumping rate, the superradiance due to the 2p−1s transitions should appear at currents lower than those needed for the 2s−2p transitions. Therefore, if this point of view is adopted, we cannot explain why superradiance is observed for transitions beginning from the $2p_2$, $2p_4$, $2p_5$, and $2p_6$ levels. It follows from the direct excitation cross sections (Table 8) that the $2p_7$, $2p_8$, $2p_9$, and $2p_{10}$ levels have excitation cross sections comparable with or even greater than the levels just mentioned. The $2p_{7,8,9,10}−1s_{2,3,4,5}$ transitions have higher values of the product $A_{ik}\lambda^3$ (see below) than the corresponding $2p_{2,4,5,6}−1s$ transitions but, in spite of this, superradiance or laser emission is not observed (under our conditions) for the $2p_{7,8,9,10}−1s$ transitions.

Thus, experimental observations are in conflict with the hypothesis of direct electron excitation of the upper levels. On the other hand, it is well known that the 2p levels of neon are largely populated, even at low current densities, because of multistage excitation from the 1s levels [68, 69, 74, 77]. It is therefore natural to assume that the inversion of the 2p−1s transitions is due to multistage excitation of the 2p levels by electrons [16]. Since superradiance due to the 2p−1s transitions is observed at the beginning of a current pulse, we shall be interested in the processes resulting in the filling of the neon levels during this stage. If the pulse repetition frequency is not too high, as in our experiments, all the neon atoms are in the ground state before each current pulse. It follows from general considerations and from Table 8 that the resonance level $1s_2$ is populated fastest and most efficiently during the initial moments of the flow of the current. The higher population of the $1s_2$ level is indicated by the observation that neither neon nor any other inert gas exhibits superradiance or laser emission terminating at this level although this should be observed first because the product $A_{ik}\lambda^3$ is greater than for any other transition (Table 10).

The excitation cross sections of the 2p levels from the $1s_2$ level are very large, of the order of 10^{-15} cm^2 (Table 9). Moreover, under conditions typical of pulse discharges in neon, the average electron energy is close to that corresponding to the maximum of the excitation cross sections of the 1s−2p transitions. Consequently, the efficiency of multistage population of the 2p levels via the $1s_2$ resonance level should be extremely high. It is clear from Table 9 that the greatest rate of the multistage population is achieved for the $2p_1$, $2p_4$, $2p_5$, and $2p_6$ levels, i.e., for those levels which are active in superradiance and laser emission. In contrast to the direct excitation process, the multistage excitation by electrons is proportional not to the first power but to the square of the electron density. This means that population inversion of the 2p−1s transitions should be achieved very readily in fast discharges and high current densities, as found experimentally. On the other hand, pulse laser emission due to the 2s−2p transitions resulting from the direct excitation of the 2s levels is observed as a result of

TABLE 10

Level	$1s_5$		$1s_4$		$1s_3$		$1s_2$	
	λ, Å	$\lambda^3 A_{ik}$, rel. units	λ, Å	$\lambda^3 A_{ik}$, rel. units	λ, Å	$\lambda^3 A_{ik}$, rel. units	λ, Å	$Q(1s_2−2p)\cdot10^{16}$, cm^2
$2p_1$			5400	11			5852	69
$2p_2$	5882	167	6029	105	6163	358	6599	17
$2p_3$			6074	1255			6652	0
$2p_4$	5945	190	6096	354			6678	23
$2p_5$	5975	66	6128	11.5	6266	558	6717	21
$2p_6$	6143	535	6305	113			6929	64
$2p_7$	6217	125	6383	787	6533	330	7024	0
$2p_8$	6334	417	6506	795			7174	14
$2p_9$	6402	1350						0
$2p_{10}$	7032	850	7245	385	7439	128	8082	0

relatively weak excitation [75]. Moreover, population inversion by multistage excitation should be delayed somewhat in time compared with the inversion due to direct population. This is in agreement with the observations [14] that the infrared superradiance is 8 nsec earlier than the visible radiation. Poor time resolution of our apparatus in the infrared part of the spectrum has prevented us from revealing this delay. However, judging by the results plotted in Fig. 51, superradiance due to the 2p−1s transitions is delayed slightly (3-5 nsec) relative to the 5350 Å thallium superradiance, which is entirely due to direct processes. This also supports the hypothesis of multistage population.

We shall first identify the lines for which we can expect superradiance as a result of multistage population. Clearly, superradiance should be observed earliest for the transitions characterized by the highest values of the gain. At a line maximum, the gain is

$$K = \frac{\lambda^2}{4} \frac{A_{ik}}{\Delta\omega} \left(N_i - \frac{g_i}{g_k} N_k \right),$$

where λ is the wavelength of the $i \rightarrow k$ transition; A_{ik} is the transition probability; $\Delta\omega$ is the line width expressed as angular frequency; N_i and N_k are the populations of the upper and lower levels, respectively; g_i and g_k are the statistical weights of the levels. In the Doppler broadening case, we find that $K \propto \lambda^3 A_{ik}$.

Table 10 gives the wavelengths and values of $\lambda^3 A_{ik}$ for all the allowed transitions from the 2p levels to the $1s_{2,3,4,5}$ levels. The values of A_{ik} are taken from [68]. It is clear that, in accordance with the assumed scheme, there should be no superradiance terminating at the $1s_2$ level. The column for the $2p-1s_2$ transitions gives the cross sections for the excitation of the 2p levels from the $1s_2$ level taken from [77]. The values underlined in Table 10 refer to the transitions for which superradiance or laser emission is observed. It is clear from the last column of Table 10 that the $2p_1$, $2p_4$, $2p_5$, and $2p_6$ levels should be populated most strongly. Superradiance from these levels should be observed for the transitions characterized by the highest values of the gain. This is in ageement with the observations that the superradiance from the $2p_6$ and $2p_5$ levels is observed as a result of transitions for which the product $\lambda^3 A_{ik}$ has the highest value. Superradiance from the $2p_4$ level is observed for the transition whose product $\lambda^3 A_{ik}$ is not the highest. This is due to the fact that the $2p_4-1s_4$ transition with a high value of the same product terminates at a level which is optically coupled to the ground state and which can be populated rapidly by electrons. The 5400 Å ($2p_1-1s_4$) line needs special consideration because its behavior differs considerably from the behavior of the other superradiance lines. This line is observed in wider tubes and in a wider range of neon pressures. The gain of this line is considerably less than the gains of the other lines. Under laser emission conditions, this line is characterized, in contrast to other lines, by a high power. It is clear from Table 10 that only one laser emission line − 5400 Å − should be emitted from the $2p_1$ level as a result of multistage excitation. The factor $\lambda^3 A_{ik}$ for this line is an order of magnitude smaller for the other superradiance lines. Consequently, the gain of this line should be, for a given pumping rate, considerably higher. Moreover, the $2p_1$ level is coupled by a strong optical transition to the $1s_2$ level and the probability of this transition is much higher than to the $1s_4$ level [68]. Therefore, efficient population inversion of the $2p_1-1s_4$ transition requires reabsorption of the emitted radiation as a result of the $2p_1-1s_2$ pumping transition, which does occur when the discharge tube diameter is large and the pressure is high.

Thus, only the multistage excitation of the 2p levels by electrons via the $1s_2$ resonance level explains the experimentally observed features of the superradiance and laser emission as a result of the 2p−1s transitions in neon.

§2. Superradiance and Laser Emission due to Transitions in Argon, Krypton, and Xenon

As mentioned earlier, the superradiance due to the 2p−1s transitions is observed not only in neon but also in krypton and xenon [17]. Superradiance lines and the corresponding transitions are listed in Table 7. Argon exhibits pulse superradiance due to other transitions [80] but not for the 2p−1s transitions. It is not clear whether this is due to a different population inversion mechanism in the case of argon or due to some special features of the population of the argon levels. We decided to determine to what extent the population inversion mechanism established for neon applied to other inert atoms with similar level structures. This was particularly important in the case of argon. With this in mind, we carried out an experimental study of the pulse superradiance emitted from argon, krypton, and xenon.

The experiments on argon, krypton, and xenon were carried out using the same apparatus as for neon. The procedure revealed easily all the superradiance lines of krypton and xenon reported in [17]. No new superradiance lines were obtained for these two gases. We therefore decided that there was little point in analyzing the experimental results obtained for krypton and xenon.

We filled a tube (1.3 mm in diameter and of active length 20 cm) with argon and observed for the first time the pulse superradiance due to the following three lines resulting from the 2p−1s transitions:

Wavelength λ, Å .	7634.8 ± 0.5	7723.8 ± 0.5	7948.0 ± 0.5
Transition	$2p_6 - 1s_5$	$2p_7 - 1s_5$ or $2p_2 - 1s_3$	$2p_4 - 1s_3$

Figure 52 shows the dependences of the energy of the superradiance pulses at the three wavelengths on the argon pressure. We found that superradiance due to the 2p−1s transitions

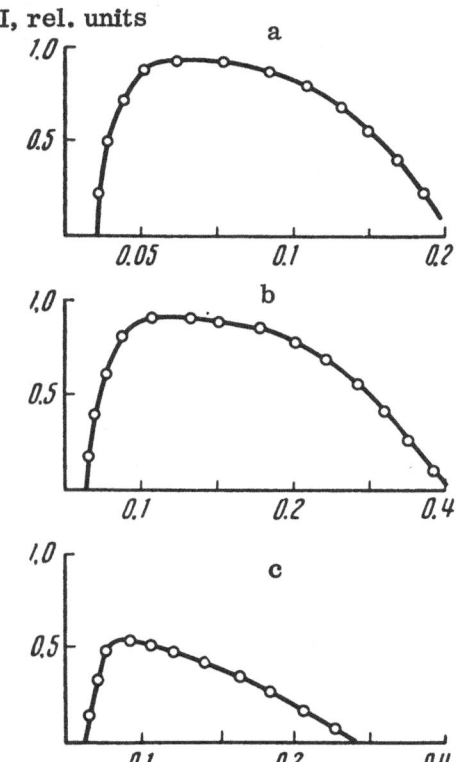

Fig. 52. Dependences of the energy of superradiance pulses representing various argon lines (2p−1s transitions) on the argon pressure: a) 7635 Å line; b) 7724 Å; c) 7948 Å.

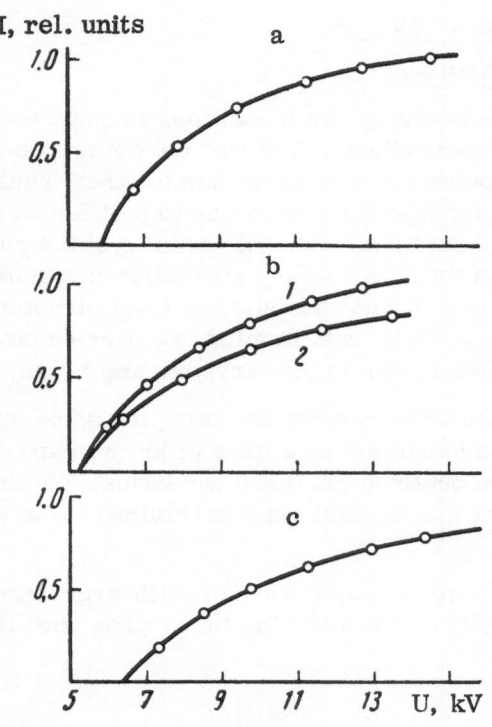

Fig. 53. Dependences of the energy of super-radiance pulses corresponding to various argon lines (2p—1s transitions) on the voltage applied to the transformer: a) 7635 Å line; b) 7724 Å line (1 — with mirror behind the tube, 2 — without mirror); c) 7948 Å.

was observed in the pressure range 0.03-0.3 Torr. The optimal pressure was 0.06-0.08 Torr. In these measurements we used one mirror, which was placed at a distance of 30 cm from the active part of the tube. Figure 53 gives the dependences of the energy of the superradiance pulses at these wavelengths on the voltage applied to the pulse cable transformer. It is clear that a mirror behind the tube had no influence on the superradiance energy at the wavelengths 7635 and 7948 Å and it only increased slighly the energy of the superradiance at 7724 Å. The nature of the dependence of the energy of the superradiance pulses on the transformer voltage was the same for argon as for neon and, in particular, a tendency to saturation was observed.

Apart from the lines listed above, argon excited in the same tube emitted a much stronger superradiance due to other transitions:

Wavelength λ, μ . . .	1.21	1.24	1.27
Transition	$3s_1'-2p_4$	$3d_2-2p_7$	$3s_1'-2p_2$

These superradiance lines were reported in [80]. Under our conditions, they were observed in the pressure range 0-20 Torr. The optimal pressure was several Torr. In contrast to the superradiance due to the 2p—1s transitions observed under the same experimental conditions, we found that the superradiance in the 1.2 μ range was strongly influenced by the presence of a mirror behind the tube. This mirror caused the energy to increase considerably (Fig. 54). As in the experiments with neon, thallium and lead, the superradiance at all the argon wavelengths was not affected when the mirror was moved to distance exceeding 1 m from the active part of the tube.

We also excited argon in a tube of 6.5 mm in diameter and 120 cm long. In this case, no superradiance was observed due to the 2p—1s transitions. However, many superradiance lines were found in the 2 μ region, but they were not investigated.

An infrared detector with a good time resolution and a sufficiently high sensitivity was not available to us during this investigation. Consequently, we were unable to record the time

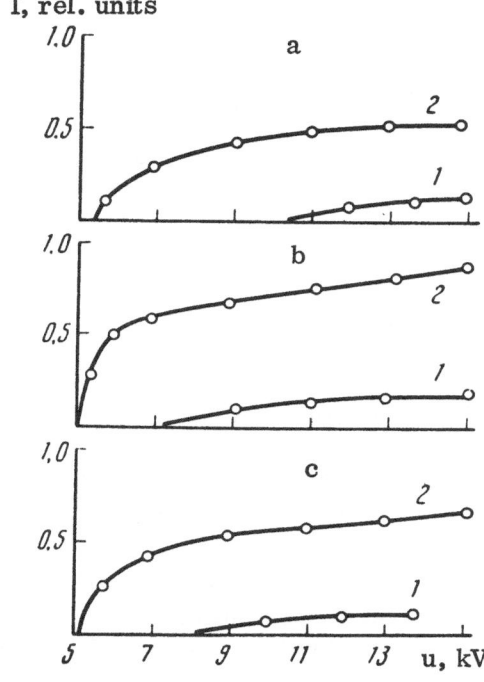

Fig. 54. Dependences of the energy of super-radiance pulses representing various infrared argon lines on the transformer voltage: a) 1.21 μ line; b) 1.24 μ; c) 1.27 μ; 1) superradiance in the absence of mirror; 2) superradiance in the presence of a mirror separated by 30 cm from the active part of the tube.

characteristics of superradiance lines of argon, krypton, and xenon. Dependences of the energy of the superradiance pulses on the distance between the mirror and the active part of the tube indicated that the duration of the superradiance pulses was of the order of a few nanoseconds and possibly less.

Before discussing the population inversion of the argon transitions, we shall first give the published information on the population inversion of the 2p−1s transitions. Apart from the lines given above, superradiance due to the 2p−1s lines of argon was reported later in [67]. The experiments reported in that paper were carried out in a tube whose internal diameter was 3 mm and whose length was 126 cm. In the pressure range 0.03-0.07 Torr, we observed superradiance at the following two wavelengths:

Wavelength λ, Å	6965.4	7067.2
Transition	$2p_2 - 1s_5$	$2p_3 - 1s_5$

Superradiance or laser emission due to the transitions observed in our study was not found in [67]. This was clearly due to different excitation conditions.

In an analysis of the population inversion processes, it is very desirable to know also the electron-impact excitation cross sections of the argon levels, similar to those available for neon. Unfortunately, very little has been published on the excitation cross sections of argon, krypton, or xenon. The most detailed information on the excitation cross sections of the 2p levels of all the inert gases is given in [81]. This paper reports the cross sections for the electron-impact excitation of the 2p levels from the ground state. The values obtained are given (in units of 10^{-19} cm^2) in Table 11.

The cross sections in this table should be regarded as approximate because of lack of allowance for the cascade transitions. These transitions may represent contributions ranging from 10 to 50%.

TABLE 11

Level	Argon	Krypton	Xenon	Level	Argon	Krypton	Xenon
$2p_1$	110	31.8	27.5	$2p_6$	129.8	102.3	100
$2p_2$	54.6	108.9	0.33	$2p_7$	100	126.3	108
$2p_3$	84.4	52.5	33.4	$2p_8$	296.8	220	226.7
$2p_4$	100	79.8	15.5	$2p_9$	195.5	111.3	240
$2p_5$	46.7	86.5	80.7	$2p_{10}$	331.5	325.5	190.6

If we assume that the population inversion is due to direct excitation of the 2p levels by excitons from the ground state, we find that the highest population should be expected for the argon levels $2p_{10}$, $2p_8$, and $2p_9$ because the excitation cross sections of these levels are largest. However, superradiance is observed from the $2p_4$, $2p_6$, and $2p_7$ levels and — under different excitation conditions — from the $2p_2$ and $2p_3$ levels. We shall show later that this discrepancy cannot be explained by the different transition probabilities. Thus, the available information does not agree with the population inversion by direct electron excitation of the 2p levels from the ground state.

We shall now consider the cascade population inversion mechanism. Obviously, as in the case of neon, stimulated cascades should play the main role. Superradiance representing lines terminating at the $2p_4$, $2p_7$, and $2p_2$ levels is observed experimentally. This superradiance may serve as the pumping radiation for lines beginning from these levels. However, the considerable difference between the dependences on the experimental conditions exhibited by the superradiance lines beginning from the 2p levels and the lines of the postulated cascade suggests that the cascade population inversion mechanism does not agree with the experimental data for argon. The difference is particularly striking between the dependences on the argon pressure and in the behavior when a mirror is placed behind a discharge tube.

We shall now see to what extent the experimental results for argon can be explained by the multistage population of the 2p levels of electrons through an intermediate resonance level. There is no published information on the excitation cross sections of the 1s levels in argon or on the cross sections for the excitation of 2p levels from the 1s levels. Therefore, in subsequent analysis of the results obtained for argon, we shall assume that the excitation cross sections of the allowed transitions are much larger than those of the forbidden transitions and that the former are proportional to the oscillator strengths. These strengths for the resonance lines of inert gas atoms are listed below (they are taken from [82]):

	$f(1s_2)$	$f(1s_4)$		$f(1s_2)$	$f(1s_4)$
Neon	0.131	0.009	Krypton . . .	0.173	0.173
Argon	0.186	0.047	Xenon	0.19	0.26

We can see that, according to our hypothesis, the rate of electron excitation of the $1s_2$ level of neon is an order of magnitude greater than the corresponding rate of excitation of the $1s_4$ level; in the case of argon, the difference is a factor of 4 whereas, in the case of krypton and xenon, the rates of excitation of these two resonance levels are approximately equal. Therefore, in the first approximation, we can assume that the multistage electron excitation of argon, like that of neon, occurs primarily through the $1s_2$ level. We shall allow for the excitation through the $1s_4$ level by assuming that the population of this level is one-quarter that of the $1s_2$ level.

In view of the absence of any information on the cross sections for the excitation of the 2p levels from the 1s levels, we shall estimate these cross sections using the approximate formula (30). Table 12 gives the wavelengths, oscillator strengths [82], and estimates of the

TABLE 12

Level	1s₂			1s₄			σ(1s₂)+1/4σ(1s₄)
	λ, Å	f_{ki}	σ_{max}	λ, Å	f_{ki}	σ_{max}	
$2p_1$	7504	0.13	1.6	6677	0.00	—	1.6
$2p_2$	8264	0.17	2.5	7273	0.02	0.2	2.55
$2p_3$	8408	0.43	6.45	7384	0,12	1.4	6,8
$2p_4$	8521	0.16	2.5	7472	—	—	2.5
$2p_5$	8578	—	—	7515	0.12	1.4	0.4
$2p_6$	9224	0.12	2.3	8006	0.07	1.0	2.5
$2p_7$	9354	0.01	0.3	8104	0.27	3.8	1.2
$2p_8$	9784	0.04	0.8	8425	0,41	6,1	2.3
$2p_{10}$	11488	0.005	0,14	9658	0.08	1,6	0.6

cross sections at the maxima (in units of 10^{-15} cm^2) for transitions from the $1s_2$ and $1s_4$ levels to various 2p levels. The last column gives the sum of the excitation cross sections, which governs the rate of multistage excitation of the 2p levels via both resonance levels.

The cross section for the excitation from the $1s_4$ level is taken with a factor of $^1/_4$ because we are assuming that the population of this resonance level is one-quarter less than that of the other level. It is clear from Table 12 that the $2p_3$, $2p_2$, $2p_6$, and $2p_4$ levels should be excited most rapidly and mainly via the $1s_2$ level. The $2p_8$ level should be excited somewhat less rapidly via $1s_2$ but it should be mainly populated via the level $1s_4$. The level $2p_9$ is not coupled by an optically allowed transition to a resonance level and, therefore, we shall assume that it is not excited by the multistage process.

We shall estimate the relative gains of the various transitions so as to find which transitions may be expected to produce superradiance first. If we postulate multistage population of the 2p levels by electrons and a low population of the lower levels, we find that the gain is proportional to the product of the pumping rate estimated above and the quantity $\lambda^3 A_{ik}$. Table 13 lists the values of A_{ik} taken from [82], $\lambda^3 A_{ik}$ and $\lambda^3 A_{ik} \Sigma \sigma_{max}$ for the transitions in question, as well as the total probabilities of the decay (deactivation) of the 2p levels and the multistage pumping rates estimated above. The values underlined in Table 13 represent transitions found experimentally which give rise to stimulated radiation, and the values identified by dotted lines correspond to transitions resulting in the 7724 Å superradiance line. It is not yet clear from the experimental results which transition is responsible for this line. We can see that stimulated radiation is observed experimentally for those transitions whose product $\lambda^3 A_{ik} \Sigma \sigma_{max}$ is largest.

TABLE 13

		$2p_1$	$2p_2$	$2p_3$	$2p_4$	$2p_5$	$2p_6$	$2p_7$	$2p_8$	$2p_{10}$
	$A_{ik}\cdot10^{-7}$, sec^{-1}	4,77	3.83	3.73	3.52	4.31	3.81	3.75	3.47	2.87
	[$\sigma_{max}(1s_2)+^1/_4\sigma_{max}(1s_4)$]$10^{16}$, cm^2	1.6	2.5	6.8	2.5	0.4	2.5	1.5	2.3	0.15
$1s_2$,	$A_{ik}\cdot10^{-7}$, cm^{-1}	4,75	1,68	2,46	1.48	—	0.59	0.12	0.16	0.02
$1s_3$	$A_{ik}\cdot10^{-7}$, sec^{-1}		1,27		1.97			0.28		0.11
	$\lambda^3 A_{ik}$		5,85		9,9			1.82		1,34
	$\lambda^3 A_{ik}\sum\sigma_{max}$		14,9		24.4			2.24		0.74
$1s_4$	$A_{ik}\cdot10^{-7}$, sec^{-1}	0,02	0.20	0.87	0.00	4.31	0,47	2.78	2.34	0.60
	$\lambda^3 A_{ik}$	0,07	3.50	0.01	18.3	2,41	14.8	14,6	0.77	5,40
	$\lambda^3 A_{ik}\sum\sigma_{max}$	0,11	1.95	23.8	0.03	6.60	6.07	18,2	32,7	2,97
$1s_5$	$A_{ik}\cdot10^{-7}$, sec^{-1}		0.68	0.39	0.06		2.75	0.57	0.97	2,13
	$\lambda^3 A_{ik}$		2.29	1.39	0,24		12.2	2,65	4,98	16,2
	$\lambda^3 A_{ik}\sum\sigma_{max}$		5.82	9,43	0.59		30.8	3.26	11.7	8.9

In accordance with the adopted point of view, it is more likely that the 7724 Å line is due to the $2p_2 - 1s_3$ transition. In our experiments, the stimulated radiation terminating at the $1s_4$ level is not observed, in spite of the fact that the product mentioned above is large for several of these transitions. This is clearly due to the fact that the $1s_4$ level of argon is filled relatively rapidly by direct transfer from the ground state.

Some comments are also due on the $2p_2 - 1s_5$ and $2p_{10} - 1s_5$ transitions. Stimulated radiation due to these transitions is not observed although the product $\lambda^3 A_{ik} \Sigma \sigma_{max}$ is larger than for the $2p_2 - 1s_5$ and $2p_3 - 1s_5$ transitions, for which superradiance is reported in [67]. As mentioned above, the $2p_8$ and $2p_{10}$ levels are populated by a multistage process mainly via the $1s_4$ level, whereas the levels exhibiting stimulated radiation are populated through the $1s_2$ level. Thus, the absence of population inversion for the transitions from the $2p_8$ and $2p_{10}$ levels may be explained if it is assumed that the relative rate of the population of the $1s_4$ level is less than that postulated in our calculations. We assumed that the rates of population of the $1s_2$ and $1s_4$ levels were proportional to the corresponding oscillator strengths. It could be that, in fact, the rate of excitation of the $1s_4$ level was less.

Thus, our analysis shows that all the experimental results obtained for argon also fit well the hypothesis of population inversion of the $2p - 1s$ transitions by multistage population of the upper levels with electrons mainly via the resonance level $1s_2$.

The experimental data on krypton and xenon are insufficient and not enough is known about their levels and transitions to analyze in a similar way the population inversion mechanism of the $2p - 1s$ transitions. By analogy with neon and argon, we may expect multistage excitation to play an important role. However, in the case of krypton and zenon, we have to allow also for the excitation via the $1s_4$ level.

CONCLUDING REMARKS

§ 1. Pulse Lasers Utilizing Transitions
from Resonance to Metastable Atomic Levels

It is shown above that lasers utilizing such transitions are most likely to give high efficiencies and high peak output powers. Table 14 summarizes all transitions of this kind for which laser emission or superradiance has been observed. The transitions with the highest efficiency are collected in Table 15. This table gives not only the characteristics of the levels but the experimentally obtained parameters of laser emission. Unfortunately, information on the electron-excitation cross sections of resonance levels of these atoms is lacking. Therefore, Table 15 gives the maximum excitation cross sections of the upper level σ_{max}, calculated using the approximate formula (30), which is valid for allowed transitions.

It is clear from Table 14 that stimulated emission has now been observed for a considerable number of transitions from resonance to metastable levels in many atoms. All the experimental observations are in good agreement with the population inversion mechanism discussed in the Introduction. The validity of the basic assumption of the smallness of the electron-excitation cross sections of metastable levels, compared with the corresponding cross sections of resonance levels, has been confirmed by many examples. Hence, we may assume that other transitions of this kind should result in a stimulated emission with similar properties.

It is clear from Table 15 that the green thallium line has the maximum efficiency. The somewhat lower ultimate efficiency η_{ult} is obtained for some transitions in lead and copper

TABLE 14

λ, Å	Atom	Transition	Reference	η_{ult}, %	T, °C
3639	Pb I	$6p7s\ ^3P_1^0 - 6p^2\ ^3P_1$	58	39	1000
3984	Hg II	$6p^2\ ^3P_{3/2}^0 - 6s^2\ ^2D_{5/2}$	59	25 i*	70
4057	Pb I	$6p7s\ ^3P_1^0 - 6p^2\ ^3P_2$	58	44	1000
4062	Pb I	$6p6d\ ^3D_1^0 - 6p^2\ ^1D_2$	58	33	1000
5105	Cu I	$4p\ ^2P_{3/2}^0 - 4s^2\ ^2D_{5/2}$	7, 11, 83, 84, 85, 86	38	1500
5341	Mn I	$y^6P_{7/2}^0 - a^6D_{9/2}$	7, 10	29	1200
5350	Tl I	$7^2S_{1/2} - 6^2P_{3/2}$	47	47	800
5420	Mn I	$y^6P_{5/2}^0 - a^6D_{7/2}$	7, 10	30	1200
5470	Mn I	$y^6P_{5/2}^0 - a^6D_{5/2}$	7, 10	25	1200
5517	Mn I	$y^6P_{3/2}^0 - a^6D_{3/2}$	7, 10	25	1200
5538	Mn I	$y^6P_{3/2}^0 - a^6D_{1/2}$	7, 10	17	1200
5782	Cu I	$4p^2P_{1/2}^0 - 4s^2\ ^2D_{3/2}$	7, 11, 83, 84, 86	38	1500
6278	Au I	$6p^2P_{1/2}^0 - 6s^2\ ^2D_{3/2}$	84	29	1500
7229	Pb I	$6p7s^3P_1^0 - 6p^2\ ^1D_2$	9, 54, 55	24i	1000
8542	Ca II	$4p^2P_{3/2}^0 - 3d^2D_{5/2}$	7	28 i	700
8662	Ca II	$4p^2P_{1/2}^0 - 3d^2D_{3/2}$	7	31 i	700
10330	Sr II	$5p^2P_{3/2}^0 - 4d^2D_{5/2}$	87	25 i	500
10918	Sr II	$5p^2P_{3/2}^0 - 4d^2D_{3/2}$	87	27 i	500
12900	Mn I	$z^6P_{7/2}^0 - a^6D_{9/2}$	7, 10	14	1200
13294	Mn I	$z^6P_{7/2}^0 - a^6D_{7/2}$	7, 10	15	1200
13319	Mn I	$z^6P_{5/2}^0 - a^6D_{7/2}$	7, 10	17	1200
13627	Mn I	$z^6P_{5/2}^0 - a^6D_{5/2}$	7, 10	15	1200
13864	Mn I	$z^6P_{3/2}^0 - a^6D_{3/2}$	7, 10	14	1200
13997	Mn I	$z^6P_{3/2}^0 - a^6D_{1/2}$	7, 10	9	1200
20583	He I	$2^1P_1 - 2^1S_0$	47	0.7	20
55460	Ca I	$4p^1P_1^0 - 3d^1D_2$	87	4.8	600
64560	Sr I	$5p^1P_1^0 - 4d^1D_2$	87	4.5	500

*The letter "i" indicates that η_{ult} was calculated for the excitation from the ground state of an ion.

TABLE 15

Atom	Transition	λ, Å	$\sigma_{max} \cdot 10^{16}$, cm^2	τ_u, nsec	τ_{st}, nsec	η_{ult}, %	Experimental results			
							P_p, W	P_p/V, W/cm^3	Δt, nsec	T, °C
Tl I	$7^2S_{1/2} - 6^2P_{3/2}$	5350	4.0	7.6	15	47	1800	1300	1—3	800
Pb I	$6p7s^3P_1^0 - 6p^2\ ^3P_2$	4057	3.7	6	9	44			1—3	1000
	$6p7s^3P_1^0 - 6p^2\ ^3P_1$	3639	3.7	6	31	39			1—3	1000
	$6p6d^3D_1^0 - 6p^2\ ^1D_2$	4062	3.8	4	9	33			3—5	1000
	$6p7s^3P_1^0 - 6p^2\ ^1D_2$	7229	3.7	6	250	24	$2 \cdot 10^3$	250	15	1030
Cu I	$4p^2P_{3/2}^0 - 4s^2\ ^2D_{5/2}$	5105	9.7	7.2	770	38	$40 \cdot 10^3$	25	16	1500
	$4p^2P_{1/2}^0 - 4s^2\ ^2D_{3/2}$	5782	4.5	7.2	370	38				
Mn I	$y^6P_{5/2}^0 - a^6D_{7/2}$	5420	2.1	9	435	30	100	1	20	1200
	$y^6P_{7/2}^0 - a^6D_{9/2}$	5341	2.0	9.6	300	29				
Ca II	$4p^2P_{1/2}^0 - 3d^2D_{3/2}$	8662	1.5	6.7	830	28	30		30	700
	$4p^2P_{3/2}^0 - 3d^2D_{5/2}$	8542	1.5	6.7	770	31				
Ne I	$2p_1 - 1s_4$	5400		14	1430	9	$190 \cdot 10^3$		1.5	20

atoms. The highest peak output power is due to the green copper line and it amounts to 40 kW [83, 84]. The highest practical stimulated emission efficiency is obtained for copper: Its value is 1.2% [83, 84]. The highest specific peak output power corresponds to the green thallium line, investigated by us, and it is 1.3 kW/cm^3. Table 15 includes, for comparison, experimental results obtained for the green neon line. We can see that, in spite of η_{ult} of this line being considerably less than for the thallium, lead, and copper lines, the peak output power emitted by neon is considerably higher: 190 kW [65]. The cause of this discrepancy is discussed later.

The results obtained in the present paper for thallium and those reported in [83, 84] for copper demonstrate that the instantaneous efficiency of the conversion of the pumping power into stimulated radiation can be quite high, amounting to several percent or more. However, the practical efficiency of laser emission for all the investigated lines is considerably lower than the ultimate efficiency because the excitation (current) pulses are considerably longer than the stimulated radiation pulses. Moreover, the peak output power is also lower than that predicted. This is likely to be due to the deficiencies of the power supply system, particularly to the insufficiently steep leading edges of the excitation pulses. Thus, in principle, pulse laser emission characterized by a high efficiency and high output power should be obtained for the transitions under consideration. However, this can be achieved only after overcoming serious difficulties, mainly of a technical nature. The principal difficulties are related to the design of a discharge cavity capable of operating at a sufficiently high temperature and of a pulse power supply system which is matched to this discharge cavity and produces sufficiently short pulses (of the order of the duration of stimulated emission) with steep leading edges.

The best practical efficiency and total power have been obtained so far for transitions in copper. The probabilities of these transitions are relatively low so that the requirements in respect of the leading edge and duration of the excitation pulses are relatively less stringent. However, the very high working temperature (1500-1600°C) complicates considerably the operation of practical devices.

The experimental conditions are much easier at temperatures below 1000°C because we can then use fused quartz (vitreous silica). Among lasers operating below 1000°C, the most promising is the one based on the green thallium line. The working temperature of this laser is about 800°C and the ultimate efficiency is 47%. The highest specific peak output power (1.3 kW/cm^3) has been reported for this laser so that an increase in the active volume should make it possible to generate high powers. However, efficient laser action in thallium requires a power supply system generating short (several nanoseconds) high-power excitation pulses.

The lead vapor laser occupies an intermediate position between the thallium and copper lasers: The working temperature for lead is about 1000°C, so that we can still use fused quartz. A special feature of the lead laser is that it emits several stimulated emission lines in different parts of the spectrum. Some of these lines can be generated only by excitation pulses with steep leading edges, whereas others can be excited by pulses with moderately steep edges.

We shall now consider other potential transitions from resonance levels, apart from those listed in Table 14. Such transitions occur in elements with partly filled p-shells and in some other elements. A list of these transitions is given in [7]. For some of them, stimulated emission has already been observed and they are included in Table 14. There are some restrictions why all the elements with a suitable level structure cannot be used. A considerable number of elements exist as stable molecules in the vapor state. Other elements are converted into atomic vapors but extremely high temperatures are needed to obtain a vapor of sufficiently high density. We shall consider only the elements which can be converted into a vapor of 0.1 Torr pressure at a temperature not exceeding 1600°C. Moreover, we shall ignore all the radioactive elements. Metastable levels of some elements have to be eliminated

TABLE 16

Atom	Transition	λ, Å	$\sigma_{max} \cdot 10^{16}$, cm²	η_{ult}, %	τ_{st}, nsec	N_u/N_0	T, °C at 0.1 Torr
Si I	$4s\,^3P^0_1 - 3p^2\,^1S_0$	4102	2	15	—	$2 \cdot 10^{-5}$	1480
	$4s\,^1P^0_1 - 3p^2\,^1S_0$	3905	—	15	34	$2 \cdot 10^{-5}$	1480
	$4s\,^3P^0_1 - 3p^2\,^1D_2$	2987	2	52	50	$2 \cdot 10^{-2}$	1480
Ge I	$5s\,^3P^0_1 - 4p^2\,^1S_0$	4685	4.2	14	200	$1 \cdot 10^{-7}$	1590
	$5s\,^3P^0_1 - 4p^2\,^1D_2$	3269		50	15	$1 \cdot 10^{-7}$	1590
Sn I	$6s\,^3P^0_1 - 5p^2\,^1S_0$	5631	3.5	13	700	10^{-5}	1410
	$6s\,^1P^0_1 - 5p^2\,^1S_0$	4524	1.6	16	25	10^{-5}	1410
	$6s\,^3P^0_1 - 5p^2\,^1D_2$	3801	3.5	47	15	$2 \cdot 10^{-3}$	1410
	$6s\,^1P^0_1 - 5p^2\,^1D_2$	3262	1.6	49	2.5	$2 \cdot 10^{-3}$	1410
Ba I	$6p\,^1P^0_1 - 5d\,^1D_2$	14999	19.5	26	—	$2 \cdot 10^{-7}$	720
	$6p\,^1P^0_1 - 5d\,^3D_2$	11303	19.5	31	—	$5 \cdot 10^{-6}$	720
Au I	$6p\,^2P^0_{3/2} - 6s^2\,^2D_3$	3122	0.5	47	20	10^{-3}	1570
Bi I	$7s\,^4P_{1/2} - 6p^3\,^2D^0_{3/2}$	4722	10	43	110	$6 \cdot 10^{-9}$	690
Cr I	$y\,^7P^0_3 - a\,^5S_2$	4942	5.5	30	2500	$5 \cdot 10^{-4}$	1510
	$y\,^7P^0_2 - a\,^5S_2$	4964	5.5	36	3600	$5 \cdot 10^{-4}$	1510
Fe I	$z\,^5F^0_5 - a\,^5F_5$	5012	1.5	37	1500	$4 \cdot 10^{-3}$	1590
	$z\,^5P^0_3 - a\,^5P_3$	8628	2.8	20	530	$6 \cdot 10^{-7}$	1590
Co I	$x\,^4D^0_{7/2} - b\,^4P_{5/2}$	4086	3.2	26	36	$4 \cdot 10^{-6}$	1310
	$z\,^4F^0_{9/2} - b\,^4P_{5/2}$	7354	20	20	6700	$4 \cdot 10^{-7}$	1310
Ni I	$z\,^3D^0_3 - b\,^1D_2$	6191	1.2	23	6000	10^{-6}	1260
	$z\,^3D^0_3 - a\,^3P_2$	7110	1.2	30	5300	10^{-7}	1260

because they are so close to the ground state that they are strongly populated at the working temperature.

When all these limitations are taken into account, we obtain a list given in Table 16, which includes transitions in atoms for which stimulated emission of the type described above can be expected. This table gives the transition, wavelength, ultimate efficiency, and temperature needed to obtain a vapor pressure of ~0.1 Torr. As in [7], this pressure is selected because it is sufficient, in most cases, for the observation of stimulated emission although the optimal pressure may be higher. Moreover, this table includes an estimate of the cross section for the excitation of the upper level from the ground state [obtained using the approximate formula (30)] and the population of the lower level at the temperature listed earlier, reduced to the population of the ground state. The transition probabilities and oscillator strengths, used in the calculation of the excitation cross sections, are taken from [57] and they should be regarded as estimates. The vapor pressures of the elements are taken from [23].

It is clear from Table 16 that the working temperature for the majority of atoms with a suitable level structure is considerably higher than 1000°C so that the difficulties mentioned above remain. The only exceptions are barium and bismuth, for which the working temperature is lower. These elements are interesting also because an estimated excitation cross section of the upper active level is relatively large. As mentioned earlier, we attempted to obtain stimulated radiation from barium and bismuth vapors. However, this attempt was unsuccessful. Since the experiments were carried out in a narrow range of experimental conditions, stimulated emission could possibly be obtained under different conditions.

It is clear from Table 16 that several silicon, germanium, tin, and gold lines have lowlying metastable levels and high ultimate efficiencies close to 50%. However, it is most difficult to achieve stimulated emission of these lines because the lower level is strongly populated at the working temperature. In the case of chromium, iron, nickel, and cobalt atoms, the level structure is very complex and it may give rise to other transitions apart from those given in Table 16. A high efficiency would be difficult to obtain for these atoms because excitation would become distributed over a large number of resonance levels. Similar comments apply also to rare-earth elements. Sufficiently high vapor pressure of some of them (Sm, Yb, Tm, Eu, La) can be reached at moderate temperatures. Many thulium, samarium, and ytterbium near-infrared stimulated emission lines are reported in [88, 89]. Some of them are clearly due to transitions from resonance metastable levels. A large number of levels of these atoms makes it difficult to achieve high efficiency and a high output power for one line.

A comparison of Tables 15 and 16 demonstrates that the discovery of new stimulated emission atomic lines of this kind can hardly alter the basic situation. In fact, the potentialities of the known stimulated emission lines of thallium, copper, and lead are not inferior to those due to transitions listed in Table 16. Moreover, new transitions will not resolve the difficulties associated with the need to work at high temperatures.

§ 2. Stimulated Emission due to Transitions in Ions

Efficient pulse lasers utilizing transitions from resonance to metastable levels in ions can be expected by analogy with the lasers utilizing the corresponding transitions in atoms. However, the analogy between the excitation processes of atomic and ionic levels is incomplete. It is true that in the electron excitation of atomic transitions one has to ionize a considerable number of atoms. Therefore, atomic and ionic levels are excited simultaneously in discharges. However, the ionic levels are always higher than the corresponding atomic levels. Therefore, if the degree of ionization is low, the energy of electrons in discharge is used mainly to excite atomic levels. Moreover, a high output power and efficiency are obtained only at sufficiently high densities of particles in the state from which pumping takes place. Therefore, if, in analogy with atoms, we consider pumping from the ground state of ions, efficient stimulated emission due to transitions from resonance to metastable levels in ions can be obtained only if the degree of ionization is nearly complete. However, even then the situation in the initial ionization of a gas may cause also direct excitation of ionic levels from the levels of neutral atoms, particularly from the atomic ground state. Metastable levels of ions may be populated by this process and this makes the conditions for a stimulated emission less favorable.

Thus, in estimating the possibility of population inversion due to transitions in ions, we have to know the distribution of the populations between the ionic levels during ionization. In most cases, it is very difficult to predict this distribution. The most favorable situation arises at a high pulse repetition frequency when a considerable number of active ions is available at the beginning of the next excitation pulse. However, it is essential that, by this moment, the population of the metastable level should decrease sufficiently. Therefore, the conditions should be such that the relaxation time of electrons is considerably longer than the metastable level lifetime. It follows that efficient emission of stimulated radiation due to transitions in ions is more difficult to obtain than in atoms.

Experimental studies of transitions from resonance to metastable levels revealed stimulated emission from calcium [7], strontium [87], and mercury [59] ions. The characteristics of the lasers utilizing these transitions are given in Table 14 and 15. Stimulated emission from calcium ions appears after a considerable delay relative to the stimulated emission from copper atoms. This is due to the fact that a finite time is necessary for the accumulation of population of the ground state of the calcium ion which would be sufficient for absorption of the resonance radiation.

In addition to the transitions listed in the preceding tables, stimulated emission of this kind can be expected also for many other lines because there are many ions with a suitable level structure. As a rule, the level structure of a singly ionized atom is similar to the struc-

TABLE 17

Ion	Transition	λ, Å	$\eta_{ult}(i)$, %	$A_{st} \cdot 10^8$, sec^{-1}	E_u, cm^{-1}	T, °C at 0.1 Torr
Ba II	$6p\,^2P^0_{3/2} - 5d\,^2D_{5/2}$	6142	44	0.095	5675	720
Ba II	$6p\,^2P^0_{1/2} - 5d\,^2D_{3/2}$	6497	51	0.075	4573	720
Ba II	$6p\,^2P^0_{3/2} - 5d\,^2D_{3/2}$	5854	39	0,012	4873	720
Pb II	$7s\,^2S_{1/2} - 6p\,^2P_{3/2}$	2203	51	2.8	14081	837
Sn II	$6s\,^2S_{1/2} - 5p\,^2P_{3/2}$	1900	61	—	4251	1410
Cu II	$4p\,^1P^0_1 - 4s\,^1D_2$	2112	40	1,2	26265	1420
Cu II	$4p\,^1P^0_1 - 4s\,^3D_1$	2015	33	—	23998	1420
Cu II	$4p\,^1P^0_1 - 4s\,^3D_2$	1970	43	—	22847	1420
Ag II	$5p\,^1P^0_1 - 5s\,^1D_2$	2280	31	—	46046	1160
Ag·II	$5p\,^1P^0_1 - 5s\,^3D_1$	2166	25	—	43739	1160
Ag II	$5p\,^1P^0_1 - 5s\,^3D_2$	2034	34	—	40741	1160
Sb II	$6s\,^3P^0_1 - 5p^2\,^3P_1$	1504	48	—	3055	400
Sb II	$6s\,^3P^0_1 - 5p^2\,^3P_2$	1565	57	—	5659	400
Sb II	$6s\,^3P^0_1 - 5p^2\,^1D_2$	1762	51	—	12790	400
Sb II	$5p^3\,^3D^0_1 - 5p^2\,^1D_2$	1869	50	—	12790	400
Bi II	$7s\,^3P^0_1 - 6p^2\,^3P_1$	1777	40	—	23324	690
Bi II	$7s\,^3P^0_1 - 6p^2\,^3P_2$	1902	47	—	17030	690
Bi II	$7s\,^3P^0_1 - 6p^2\,^1D_2$	2803	52	—	33936	690
Bi II	$6d\,^3D^0_1 - 6p^2\,^1D_2$	2143	56	—	33936	690

ture of levels of a neutral atom preceding it in the Mendeleev periodic table. This analogy can be used to suggest a large number of suitable transitions in ions.

Table 17 lists the transitions from resonance to metastable levels in ions as a result of which we may expect, by analogy with atoms, efficient stimulated emission. This list is far from complete. Many of the lines are of interest even when stimulated emission is not very efficient because they suggest possible ways of achieving stimulated emission at wavelengths shorter than 2300 Å.

§ 3. Pulse Laser Emission and Superradiance due to

2p – 1s Transitions in Inert Gases

Pulse laser emission and superradiance due to transitions of this kind were reported (subsequent to our investigation) for all the inert gases with level structures similar to that of neon. The transitions for which laser emission or superradiance was observed experimentally and some of their characteristics were listed in Table 7. It is clear from this table that the ultimate efficiency of all these transitions was considerably lower than for transitions from the first resonance level. We used neon and argon as examples that the main population inversion mechanism of these transitions was multistage excitation of the upper levels by electrons via the resonance level. In this case, the pumping efficiency would be governed by the pumping of the resonance level because the transfer from the resonance level should be most efficient (§ 1 in Chap. IV). Bearing in mind the pumping mechanism, ultimate efficiencies of the transitions, and the fact that the pumping would be spread over many levels, we concluded that the peak power and efficiency as a result of these transitions in inert gases should be considerably lower than for transitions from the first resonance level. Nevertheless, pulse laser emission as a result of these transitions in inert gases could be of considerable interest because it would be much easier to achieve in practice than in atoms characterized by high working temperatures of their vapors.

The greatest interest attaches to the stimulated emission as a result of transitions in neon because such transitions give rise to visible lines. We recall that the green neon line is characterized by the highest peak power (190 kW) in the visible part of the spectrum. This is obtained using a high-power transverse-discharge pulse supply system. A comparison of the characteristics of the green neon line with the thallium, copper, and lead lines mentioned earlier and with the results obtained for the inert gas atoms shows that a transverse discharge is to be recommended for a high peak output power. If technical difficulties associated with a high working temperature can be overcome for thallium, copper, and lead and transverse-discharge systems are used, it should be possible to reach a peak output power (for a sufficiently large active medium) considerably greater than for neon and characterized by a higher efficiency.

By analogy with pulse population inversion due to transitions in neon and argon, we can identify other systems in which the same population inversion mechanism should apply. In particular, a similar mechanism may give rise to population inversion at the 5461 Å and 4047 Å mercury lines as a result of multistage pumping via a triplet resonance level. An attempt to observe stimulated emission of these lines, carried out in the present study, was not successful. A level structure similar to that of neon and other inert gases can be found also in some singly ionized atoms of alkali metals. Transitions between these levels lie in the ultraviolet part of the spectrum. The present authors are not aware of any attempts to obtain a pulse population inversion of these transitions.

§ 4. Conclusions

1. An analysis was made of the properties of systems and characteristics of transitions as a result of which pulse stimulated radiation could be generated with a high efficiency giving rise to a high specific peak power. It was found that the most promising transitions were those from the first resonance to metastable levels. An analysis was followed by a search for suitable transitions as a result of which pulse stimulated radiation was observed for the first time from the following active media and transitions (including those with a high ultimate efficiency): 20,581 Å helium line, $\eta_{ult} = 0.7\%$; 5350 Å thallium line, 47%; 3639, 4057, and 4062 Å lead lines with ultimate efficiencies of 39, 44, and 33%, respectively; 3984 Å mercury ion line, 25%.

2. A detailed experimental study was made of the properties of these transitions, which established that all the experimental observations were in agreement with the hypothesis of direct excitation of the resonance levels by electrons and a low rate of excitation of the metastable levels (compared with the resonance levels).

3. The green thallium line, exhibiting the highest ultimate efficiency, was found to be characterized by a peak power of 1.8 kW and the highest (for all pulse lasers utilizing transitions in atoms and ions) specific peak power amounting to 1.3 kW/cm^3. The 7229 Å lead line was characterized by a peak power of about 2 kW from an active volume smaller than in earlier investigations.

4. The efficiency of conversion of the pumping power into stimulated 5350 Å thallium line was about 1% and the practical efficiency of these laser systems was considerably lower than the attainable value because the stimulated radiation pulses were much shorter than the excitation (current) pulses. An improvement in the characteristics of the power supply systems of lasers utilizing transitions from resonance to metastable levels should increase considerably the practical efficiency and the specific output power.

5. An experimental study was made of the pulse laser emission and superradiance due to 2p—1s transitions in neon, argon, krypton, and xenon. Stimulated radiation was observed for the first time due to such transitions in argon and one new transition in neon. The examples of neon and argon were used to show that the principal inversion mechanism of these transitions was the multistage excitation of the upper active levels by electrons via the intermediate resonance level.

6. Pulse superradiance at the wavelength of the green thallium line in thallium iodide vapor was observed for the first time and investigated. In this case, the inversion was due to a new mechanism involving the dissociation of the molecules by electron impact with a preferential population of one of the levels of the dissociation product (in this case, the thallium atom).

7. A detailed experimental investigation was made of the time characteristics of the stimulated radiation pulses and excitation (current) pulses applied to lasers utilizing thallium and lead vapors and neon transitions. The shape and duration of the stimulated radiation pulses depended on the discharge conditions and, in the presence of a mirror, on the distance between this mirror in the active medium. Under certain experimental conditions, short stimulated radiation pulses of about 1 nsec duration could be emitted by all the investigated lasers. These short pulses could be used in checking and calibration of fast-response photodetectors (photomultipliers, coaxial photocells, etc.).

8. A theoretical calculation was made of the saturated power of stimulated radiation emitted from a three-level system on the assumption of a linear rise of the electron density with time. The influence of the spontaneous decay and interaction with electrons in the active channel was taken into account. It was found that the peak power and duration of stimulated radiation pulses were governed by the rate of rise of the electron density. A calculation was made of the saturated stimulated radiation power carried by the green thallium line, which indicated that the specific peak power could reach 10 kW/cm^3 and that the duration of the stimulated radiation pulses was governed by the interaction with electrons.

9. The experimental results obtained in the present investigation and the results reported by other workers were used in an analysis of other possible pulse lasers of this type. Indentification was made of the active media and transitions as a result of which it should be possible to obtain stimulated radiation by population inversion mechanisms investigated in the present study. Recommendations were made on further research on lasers of this type.

The authors are deeply grateful to M. A. Kazaryan and P. I. Ishchenko for their participation in this investigation, and to all the colleagues and technicians at the Lebedev Institute, without whose help many of the experiments reported above would have been difficult to carry out.

LITERATURE CITED

1. W. R. Bennett, Jr., Appl. Opt. Suppl. on Chemical Lasers, No. 2, 3 (1965).
2. G. Gould, Appl. Opt. Suppl. on Chemical Lasers, No. 2, 59 (1965).
3. V. P. Tychinskii, Usp. Fiz. Nauk, 91:389 (1967).
4. N. N. Sobolev and V. V. Sokovnikov, Usp. Fiz. Nauk, 91:425 (1967).
5. G. N. Rokhlin, Gas-Discharge Light Sources [in Russian], Energiya, Moscow-Leningrad (1966).
6. C. Kenty, J. Appl. Phys., 21:1309 (1950).
7. W. T. Walter, N. Solimene, M. Piltch, and G. Gould, IEEE J. Quantum Electron., QE-2:474 (1966).
8. G. G. Petrash, Usp. Fiz. Nauk, 105:645 (1971).
9. G. R. Fowles and W. T. Silfuast, Appl. Phys. Lett., 6:236 (1965).
10. M. Piltch, W. T. Walter, N. Solimene, G. Gould, and W. R. Bennett, Jr., Appl. Phys. Lett., 7:309 (1965).
11. W. T. Walter, M. Piltch, N. Solimene, and G. Gould, Bull. Am. Phys. Soc., 11:113 (1966).

12.   M. Piltch and G. Gould, Rev. Sci. Instrum., 37:925 (1966).
13.   D. Rosenberger, Phys. Lett., 13:228 (1964).
14.   D. M. Clunie, R. S. A.Thorn, and K. E. Trezise, Phys. Lett., 14:28 (1965).
15.   D. A. Leonard, R. A. Neal, and E. T. Gerry, Appl. Phys. Lett., 7:175 (1965).
16.   I. N. Knyazev and G. G. Petrash, Zh. Prikl. Spektrosk., 4:560 (1966).
17.   D. Rosenberger, Phys. Lett., 14:32 (1965).
18.   E. T. Gerry, Appl. Phys. Lett., 7:6 (1965).
19.   A. W. Ali, A. C. Kolb, and A. D. Anderson, Appl. Opt., 6:2115 (1967).
20.   T. Holstein, Phys. Rev., 72:1212 (1947); 83:1159 (1951).
21.   N. P. Penkin and L. N. Shabanova, Opt. Spektrosk., 14:167 (1963).
22.   I. I. Sobelman, Introduction to the Theory of Atomic Spectra, Pergamon Press, Oxford
       (1973).
23.   A. N. Nesmeyanov, Vapor Pressure of the Chemical Elements, American Elsevier, New
       York (1963).
24.   A. S. Nasibov, A. A. Isaev, V. M. Kaslin, and G. G. Petrash, Prib. Tekh. Eksp., No. 4,
       232 (1967).
25.   V. A. Burmakin, A. A. Doroshkin, and G. G. Petrash, Elektron. Tekh. Ser 1, No. 2,
       142 (1970).
26.   G. A. Mesyats, A. S. Nasibov, and V. V. Kremnev, Formation of Nanosecond Pulses
       [in Russian], Énergiya, Moscow (1970).
27.   R. Huddlestone and S. L. Leonard (eds.), Plasma Diagnostic Techniques, Academic Press,
       New York (1965).
28.   H. S. W. Massey and B. L. Moiseiwitsch, Proc. R. Soc. A, 258:147 (1960).
29.   F. T. Arecchi, in: Quantum Electronics (Proc. Third Intern. Congress, Paris, 1963),
       Vol. 1, Dunod, Paris, and Columbia University Press, New York (1964), p. 547; Alta Freq.,
       31:722 (1962).
30.   I. P. Zapesochnyi and P. V. Fel'tsan, Ukr. Fiz. Zh., 10:1197 (1965).
31.   P. I. Ishchenko, Diploma Thesis [in Russian], Moscow Physicotechnical Institute (1966).
32.   A. H. Gabriel and D. W. O. Heddle, Proc. R. Soc. A, 258:123 (1960).
33.   R. M. St. John, F. L. Miller, and C. C. Lin, Phys. Rev., 134:A888 (1964).
34.   H. R. Moustafa Moussa, F. J. de Heer, and J. Schutten, Physiéa (Utr.), 40:517 (1969).
35.   J. D. Jobe and R. M. St. John, Phys. Rev., 164:117 (1967).
36.   G. J. Schulz and R. E. Fox, Phys. Rev., 106:1179 (1957).
37.   R. J. Fleming and G. S. Higginson, Proc. Phys. Soc. Lond., 84:531 (1964).
38.   H. K. Holt and R. Krotkov, Phys. Rev., 144:82 (1966).
39.   J. D. Jobe, J. Walker, and R. M. St. John, Bull. Am. Phys. Soc., 12:187 (1967).
40.   V. Cermak, J. Chem. Phys., 44:3774 (1966).
41.   J. L. G. Dugan, H. L. Richards, and E. E. Muschlitz, Jr., J. Chem. Phys., 46:346 (1967).
42.   V. I. Ochkur and V. F. Brattsev, Opt. Spektrosk., 19:490 (1965).
43.   B. L. Moiseiwitsch and S. J. Smith, Rev. Mod. Phys., 40:238 (1968).
44.   C. E. Moore, Atomic Energy Levels as Derived from the Analyses of Optical Spectra,
       Nat. Bur. Stand. Circ. No. 467, Vol. 1 (1949); Vol. 2 (1952); Vol. 3 (1958).
45.   S. É. Frish, Optical Spectra of Atoms [in Russian], Fizmatgiz, Moscow–Leningrad (1963).
46.   A. I. Odintsov, Opt. Spektrosk., 9:75 (1960).
47.   A. A. Isaev, P. I. Ishchenko, and G. G. Petrash, ZhETF Pis'ma Red., 6:619 (1967).
48.   A. A. Isaev and G. G. Petrash, ZhETF Pis'ma Red., 7:204 (1968).
49.   Chemist's Handbook [in Russian], Énergiya, Leningrad (1962).
50.   A. N. Terenin, Phys. Z. Sowjetunion, 2:377 (1932).
51.   Données Spectroscopiques Concernant Les Molécules Diatomiques, Hermann, Paris (1952).
52.   N. M. Frank, Phys. Z. Sowjetunion, 2:319 (1932); Tr. Gos. Opt. Inst., No. 87 (1933).
53.   V. A. Dudkin, T. L. Andreeva, V. I. Malyshev, and V. N. Sorokin, Opt. Spektrosk.,
       19:177 (1965).

54. W. T. Silfvast and J. S. Deech, Appl. Phys. Lett., 11:97 (1967).
55. J. S. Deech, J. B. Cole, and J. H. Sanders, J. Phys. B, 2:47 (1969).
56. N. P. Penkin and Yu. Yu. Slavenas, Opt. Spektrosk., 15:154 (1963).
57. C. H. Corliss and W. R. Bozman, "Experimental transition probabilities for spectral lines of seventy elements derived from NBS tables of spectral-line intensities," Nat. Bur. Stand. Monogr., No. 53, pp. III-XVII (1962).
58. A. A. Isaev and G. G. Petrash, ZhETF Pis'ma Red., 10:188 (1969).
59. A. A. Isaev and G. G. Petrash, Zh. Prikl. Spektrosk., 12:1118 (1970).
60. B. Venkatesachar and L. Sibaiya, Proc. Indian Acad. Sci. A, 1:8 (1934).
61. S. Mrozowski, Phys. Rev., 57:207 (1940).
62. S. Mrozowski, Phys. Rev., 61:605 (1942).
63. W. T. Walter, Bull. Am. Phys. Soc., 12:90 (1967).
64. D. A. Leonard, IEEE J. Quantum Electron., QE-3:133 (1967).
65. J. D. Shipman, Jr., Appl. Phys. Lett., 10:3 (1967).
66. A. A. Isaev and G. G. Petrash, Elektron. Tekh. Ser. 3, No. 3, 17 (1967).
67. V. A. Tolkachev, Zh. Prikl. Spektrosk., 8:746 (1968).
68. W. R. Bennett, Jr., and P. J. Kindlmann, Phys. Rev., 149:38 (1966).
69. S. É. Frish and V. F. Reval'd, Opt. Spektrosk., 15:726 (1963).
70. P. V. Fel'tsan, I. P. Zapesochnyi, and M. M. Povch, Ukr. Fiz. Zh., 11:1222 (1966).
71. W. R. Bennett, Jr., in: Advances in Quantum Electronics (Proc. Second Intern. Conf., Berkeley, Calif., 1961), Columbia University Press, New York (1961), p. 28.
72. P. K. Tien, D. MacNair, and H. L. Hodges, Phys. Rev. Lett., 12:30 (1964).
73. L. A. Vainshtein and L. A. Minaeva, Preprint No. 35 [in Russian], Lebedev Physics Institute, Academy of Sciences of the USSR, Moscow (1967).
74. W. R. Bennett, Jr., Appl. Opt. Suppl. on Optical Masers, No. 1, 24 (1962).
75. G. G. Petrash and I. N. Knyazev, Zh. Eksp. Teor. Fiz., 45:833 (1963).
76. C. K. N. Patel, J. Appl. Phys., 33:3194 (1962).
77. I. M. Beterov and V. P. Chebotaev, Opt. Spektrosk., 23:854 (1967).
78. R. der Agobian, J. L. Otto, R. Cagnard, and R. Echard, C. R. Acad. Sci. (Paris), 259:85 (1964).
79. O. Andrade, M. Gallardo, and K. Bockasten, Appl. Phys. Lett., 11:99 (1967).
80. K. Bockasten, T. Lundholm, and O. Andrade, Phys. Lett., 22:145 (1966).
81. P. V. Fel'tsan and I. P. Zapesochnyi, Ukr. Fiz. Zh., 12:633 (1967); P. V. Fel'tsan, Ukr. Fiz. Zh., 12:1424 (1967); P. V. Fel'tsan and I. P. Zapesochnyi, 13:205 (1968).
82. W. L. Wiese, Proc. Eighth Intern. Conf. on Phenomena in Ionized Gases, Vienna, 1967, Contributed Papers, publ. by International Atomic Energy Agency, Vienna (1968), p. 447.
83. W. T. Walter, Bull. Am. Phys. Soc., 12:90 (1967).
84. W. T. Walter, IEEE J. Quantum Electron., QE-4:355 (1968).
85. D. A. Leonard, IEEE J. Quantum Electron., QE-3:133 (1967).
86. J. F. Asmus and N. K. Moncur, Appl. Phys. Lett., 13:384 (1968).
87. J. S. Deech and J. H. Sanders, IEEE J. Quantum Electron., QE-4:474 (1968).
88. P. Cahuzac, Phys. Lett. A, 27:473 (1968).
89. P. Cahuzac and J. Brochard, J. Phys. (Paris) Suppl., 30(C1):81 (1969).

PULSE GAS LASERS UTILIZING ELECTRONIC TRANSITIONS IN DIATOMIC MOLECULES

V. M. Kaslin and G. G. Petrash

A detailed experimental investigation was made of the most important laser systems utilizing electronic–vibrational–rotational transitions in diatomic molecules. A study was made of $C^3\Pi_u - B^3\Pi_g$ (second positive band system) and $B^3\Pi_g - A^3\Sigma_u^+$ (first positive system) transitions in the N_2 molecule, $B_1\Sigma^+ - A^1\Pi$ transition (Ångstrom band system) in the CO molecule, and $2s\sigma E^1\Sigma_g^+ - 2p\sigma B^1\Sigma_g^+$ transition in the H_2 molecule. It was found that all these systems shared a common population inversion mechanism. This mechanism involved the excitation of the active levels by direct electron impact from the ground state of the molecules. The distribution of the excitation and gain between the bands was governed by the Franck–Condon principle. The energy and time characteristics of stimulated emission from tubes of different diameters were governed uniquely by the ratio E/N of the electric field E in the discharge tube to the active gas density N. Cooling of the active substance in such lasers increased strongly their gain. This made it possible to obtain a record output power from the majority of the investigated laser systems, to observe superradiance, to detect inversion of the sequence of intensities in the molecular spectra, and to reveal 180 new stimulated emission lines.

INTRODUCTION

§ 1. Characteristics of Lasers Utilizing Electronic Transitions in Molecules

This paper reports the results of an experimental investigation of the properties of pulse gas-discharge lasers utilizing electronic transitions in diatomic molecules. These results have helped in the understanding of the physical mechanisms of population inversion and in identifying the optimal operating conditions. Lasers of this kind represent one of the most important classes of gas lasers. Utilization of electronic transitions in molecules makes it possible to achieve pulse laser emission for a very large number of transitions in a very wide part of the spectrum ranging from middle infrared to far ultraviolet. The total number of stimulated emission lines observed so far is close to a thousand. It is particularly important to stress that the utilization of electronic transitions in molecules provides new opportunities for stimulated emission at short wavelengths. Laser emission at the shortest wavelength reported so far (1160 Å) has been achieved in this type of laser. This part of the spectrum is outside the operating range of other lasers.

Another important aspect is the relatively high peak power which can be obtained from laser systems of this type. For example, a laser utilizing transitions in the nitrogen molecule can emit ultraviolet power of 24 MW, which is a record for all gas lasers. Transitions in the hydrogen molecule in the vacuum ultraviolet range have given rise to an output power of the order of 1 MW but this value should be exceeded comfortably in future. We shall show that considerable peak output powers can be expected also for other systems of this kind.

The extensive range of wavelengths and the high peak powers attainable from lasers utilizing electronic transitions in molecules should be useful in applications of these lasers in science, technology, and the national economy. A pulse laser utilizing ultraviolet transitions in nitrogen is already employed quite extensively, particularly in the pumping of dye lasers, in studies of the composition of the atmosphere, and some other applications. It is planned to manufacture this laser industrially.

However, the potentialities of gas lasers utilizing electronic transitions in molecules are not yet fully realized. We can expect further extension in the direction of vacuum ultraviolet, considerable improvement in the efficiency and in the peak and average output powers, as well as continuous operation. Undoubtedly, the list of active molecules and stimulated emission lines will increase because only the simplest systems with the best known molecules (nitrogen, hydrogen, and carbon monoxide) have been made to work. Thus, the prospects for gas lasers utilizing electronic transitions in molecules look very promising.

The discharge conditions under which the lasers under discussions are operated are very close to those used in numerous and well-known lasers utilizing atomic transitions. Therefore, it is natural to expect similar population inversion mechanisms and similar properties of stimulated radiation. However, we shall show that the population inversion mechanisms of molecular transitions can hardly ever be reduced to mechanisms known for atomic systems. This is due to the fact that the excitation and deactivation of molecular levels are characterized by several specific features. The main feature of the molecules is the large number of sublevels of each electronic state and these are due to the vibrational and rotational motion of the nuclei in the molecule. Consequently, excited molecules are practically always distributed over a large number of levels. Since the excitation cross sections and deactivation rates of electronic molecular states are of the same order of magnitude as those of atomic levels, under comparable conditions each individual rotational sublevel is populated much less strongly than a corresponding atomic level of the same energy. Since this applied to both upper and lower active levels, it should not, in principle, interfere with population inversion but the gain of each rotational transition is found to be considerably less than for the corresponding atomic transition with a similar level distribution. In the case of molecules composed of atoms of moderate mass (N_2, O_2, CO) a typical population of an electronic state is distributed, roughly speaking, between 3–5 vibrational levels and each vibrational level has 10–30 rotational sublevels. Thus, the gain for a molecular transition is typically one or two orders of magnitude lower than the gain of an atomic transition.

Hence, it is clear that any means of increasing the gain are particularly important in molecular systems. This is especially true of pulse lasers in which the inverted state exists for a limited and frequently very short time. In this situation even a strong population inversion may not ensure a high stimulated emission power and efficiency if the gain is not sufficiently large. This is due to the fact that under low-gain conditions the field in a laser cannot rise, during the short inversion lifetime, sufficiently to reach saturation and utilize fully the population inversion. We shall show that a large increase in the gain and, consequently, in the efficiency and peak output power can be achieved by cooling the active gas. Therefore, a considerable attention will be paid to the influence of the gas temperature on the properties of pulse stimulated emission due to electronic transitions in molecules. Cooling also provides an additional method for identifying population inversion mechanisms in molecular lasers and for finding means for improving their characteristics.

The presence of a large number of energy levels in molecules also has a positive aspect since it provides additional opportunities for selective population and relaxation of levels and, consequently, additional opportunities for population inversion. A particularly important aspect is that the optical transitions and electron excitation obey the Franck-Condon principle. This makes it possible to use population inversion mechanisms without analogs in atomic systems.

Another special feature of lasers utilizing electronic transitions in molecules is usually a rich spectral composition of the stimulated emission. This naturally gives rise to certain difficulties in studies of the stimulated emission spectra but it also gives considerable information which can be used to draw certain conclusions on the level excitation and population inversion mechanisms. However, relatively little is known about the relationship between the population inversion in gas lasers utilizing electronic transitions in molecules and the stimulated emission spectra. In view of this a considerable attention was paid in our study to the stimulated emission spectra and their changes with the working conditions.

§ 2. Review of the Literature and Formulation

of the Problem

1. Laser emission due to electronic transitions in molecules was first reported by Mathias and Parker in 1963 for the first positive (1^+) system of the nitrogen bands in the near infrared [1]. In the same year Mathias and Parker reported pulse laser emission in the visible part of the spectrum due to the Ångstrom bands of the CO molecule [2], and Heard [3, 4] achieved laser emission in the ultraviolet part of the spectrum due to the second positive (2^+) system of the nitrogen bands. Population inversion mechanisms responsible for all these transitions were not identified. Only Mathias and Parker [1] put forward the hypothesis about the population inversion responsible for the 1^+ nitrogen system. They assumed that the direct excitation of the upper laser level $B^3\Pi_g$ from the ground state of the molecule was unlikely and they suggested that electron impacts excited the singlet state $a^1\Pi_g$ from which energy was transferred by collisions of the second kind to the upper laser state $B^3\Pi_g$.

Soon after these reports of the stimulated emission from the N_2 and CO molecules, Bazhulin et al. achieved (in 1964) laser emission due to the $E^1\Sigma_g^+ - B^1\Sigma_u^+$ transition in the hydrogen molecule [5]. Bazhulin et al. proposed a general population inversion mechanism involving direct electron-impact excitation from the ground state in which the preferential population of the upper active levels resulted from the Franck-Condon principle. A more detailed investigation of the stimulated emission spectrum, including laser emission from the D_2 and HD isotopic molecules [6, 7], confirmed the mechanism proposed in [5].

Petrash [8] analyzed the available information on the properties of stimulated emission due to electronic transitions in molecules and showed that all the results available at the time were in agreement with the mechanism described above. This mechanism served as the basis of the predictions [8, 9] of population inversion for new transitions in molecules, particularly those resulting in emission of vacuum ultraviolet radiation.

Huber [10] reported stimulated emission due to transitions in nitric oxide molecules. Investigations of stimulated emission of the CO Ångstrom bands and of the N_2 1^+ system were reported in [11-13] and some indirect information on population inversion mechanisms was obtained.

This was the situation at the time we started the present investigation of gas lasers utilizing electronic transitions in molecules. The current results and ideas on the population inversion mechanisms lead us to expect stimulated emission in a wide spectral range and with high peak output powers. However, the lack of the experimental data meant that the population inversion mechanism in the majority of the investigated lasers could not be regarded as sufficiently reliably established. Consequently, it was difficult to suggest ways of improving the properties of lasers of this kind. Moreover, reliable predictions on possible ultimate parameters of these lasers could not be made. Finally, information on the properties of these lasers and conditions for achieving optimal operation were needed for the further development of these lasers and their practical applications.

In view of this situation we decided to carry out a detailed experimental study of the properties of various lasers utilizing electronic transitions in molecules, find optimal conditions for the operation of such lasers, determine other characteristics, and then draw conclusions on the population inversion mechanisms and ultimate characteristics of lasers of this kind. This investigation continued parallel with studies of other authors. As the information was accumulated, the specific tasks in studies of each laser system were modified. A detailed formulation of our intended investigations will be given later separately for each system. The general aspects concerning all systems included the influence of temperature on the properties of stimulated radiation and studies of the stimulated emission spectra.

We studied to a greater or lesser extent all the lasers utilizing electronic transitions in molecules and emitting in the near infrared, visible, and ultraviolet parts of the spectrum, which were known at the time we began our study.

We shall now give a fuller review of the literature on each of the investigated lasers and we shall consider the tasks we set ourselves for each system.

2. Ultraviolet laser emission of the second positive system of bands of molecular nitrogen due to the $C^3\Pi_u - B^3\Pi_g$ transitions were discovered in 1963 by Heard. The first papers were very brief [3, 4]. Heard reported 30 lines ranging from 3000 to 4000 Å and emitted from pulse discharges in nitrogen. The strongest line was observed at 3371 Å. Heard concluded that the stimulated emission was due to transitions in the second positive (2^+) system of nitrogen and that the strongest line was due to the 0−0 vibrational transition in this system. An additional line at 3400 Å was described as "the brightest visible line." No other information on the spectrum was given. These results were obtained by exciting a gas with submicrosecond current pulses generated by the application of very high voltages (100−150 kV). The duration of the output radiation pulses was ~20 nsec and the total peak power was of the order of 10 W (at that time this was a high value). No conclusions were drawn in [3, 4] on the stimulated emission mechanism.

Petrash [8] analyzed all the lasers known at the time and showed that the properties of all of them agreed well with the hypothesis of direct excitation of the active levels by electron impact from the ground state of the molecule subject to the Franck−Condon principle. The lack of data on the stimulated emission of the 2^+ nitrogen system [3, 4] made it impossible to draw a reliable conclusion on the population inversion mechanism in this system. Petrash [8] suggested that the inversion in the 2^+ system was due to the same mechanism. An analysis of the Franck-Condon factors indicated that stimulated emission could be expected primarily for the 0-0 and 0-1 bands, and with greater difficulties also in the 1-0 band.

Thus, at the time we started our study there was no reliable information on this type of stimulated emission apart from its occurrence. The reports given in [3, 4] were insufficient to form any definite ideas on the properties of such stimulated emission, conditions under which it existed, or its mechanism. In particular, it was not clear whether such stimulated emission could be obtained at lower voltages. This is of practical importance because the use of voltages of 100-150 kV is highly inconvenient.

Naturally, the main problem in the study of the ultraviolet stimulated emission from nitrogen was its mechanism. This included reliable attribution of the stimulated emission lines to specific transitions without which it was not possible to consider the population inversion mechanism. Therefore, reliable spectral measurements were needed. The data given in [3, 4] were unhelpful because it was not clear to what transitions one could ascribe lines in such a wide spectral range as 3000 to 4000 Å, mentioned in these papers.

In view of this situation we started by trying to achieve ultraviolet stimulated emission of the nitrogen band at relatively low voltages and then study the properties of such emission. The next stage was a detailed investigation of the stimulated emission spectrum. We then planned to use this information to draw some conclusions on the population inversion mechanism in the new system under consideration.

The first investigation of the stimulated emission spectrum was reported by the present authors in [14], where we gave accurate stimulated emission wavelengths and attributed the various lines to definite transitions in the $0-0$ and $0-1$ bands of the 2^+ nitrogen system. Somewhat later [15, 16] we cooled the laser and found additionally stimulated emission of the $1-0$ band of the same system. The stimulated emission in the $0-0$ band was also investigated in [17-19]. A comparison of the results reported in [17-19] with our values will be made in Chap. III, § 4.

These investigations of the spectra [14-19] and other properties of stimulated emission [15-17, 20-22], as well as the calculations reported in [23-25], enabled us to justify the population inversion mechanism proposed in [8]. We then concentrated our attention on the influence of temperature on the properties of stimulated radiation.

Much work has been done on the ultraviolet stimulated emission of the nitrogen bands. The transverse discharge technique made it possible to achieve a peak output power of 200 kW in the form of pulses of 20 nsec duration. This was the method used by the American aviation firm AVCO in developing an industrial laser with a peak output power of 100 kW and a repetition frequency of 100 Hz [26]. An even higher power, up to 3 MW [27], was obtained when active gas was excited by a pulse system based on the Blumlein charging line [28]. This output power has been a record for all gas lasers until recently when an output power of 24 MW was achieved for the same system [29].

Investigations of various laser properties (maximum repetition frequency, beam divergence, mode structure of output radiation, possibility of using gas pressures up to several atmospheres, etc.) were reported in [30-31] and developments in the excitation techniques were described in [20, 27, 37-41]. In particular, the use of a pulse cable transformer in the power supply systems of gas lasers was considered in [39].

The high peak power and large gain, short duration of the output pulses, important spectral range, and other features of the ultraviolet nitrogen laser have been used in many applications [42-51]. In particular, this laser has been employed in the excitation of dye lasers, in laser ranging, communications with space vehicles, studies of the upper atmosphere, laser radar, and chemistry.

3. Pulse stimulated emission of infrared (0.75-1.23 μ) bands of the first positive system of the molecular nitrogen due to the $B^3\Pi_g - A^3\Sigma_u^+$ transitions was first achieved by Mathias and Parker in 1963 [1]. They reported that a high-voltage (20-40 kV) pulse discharge through nitrogen generated stimulated emission of about 30 lines. These lines formed four groups near the wavelengths 0.87, 0.89, 1.05, and 1.23 μ. Mathias and Parker attributed the observed lines to the $2-1$, $1-0$, $0-0$, and $0-1$ bands in the first positive (1^+) system of nitrogen. Moreover, they observed also two additional groups of lines near 0.76 and 0.77 μ, which they attributed to the $3-1$ and $2-0$ bands in the same molecular system. The total peak output power was ~100 W. Mathias and Parker also suggested a population inversion

mechanism. Since the upper laser level $B^3\Pi_g$ was a triplet, they considered it unlikely that it would be excited directly from the ground (singlet) state $X^1\Sigma_g^+$, which would involve a change in the multiplicity. It seemed more likely that the upper laser level would be populated by an indirect process as follows: Electron impact was postulated to excite the singlet $a^1\Pi_g$ from the ground state of the molecule and hence the pump energy was transferred by collisions of the second kind to the $B^3\Pi_g$. In other words, they proposed a scheme analogous to that established for the helium—neon laser.

However, it was shown in [8, 52] that the properties of such infrared stimulated emission could be explained satisfactorily by direct electron impact. A suggestion that the excitation could be due to fast electrons was also made in [12].

The other investigations of the stimulated emission due to transitions in the 1^+ system were concerned mainly with the spectrum of such emission [17, 53-56] and the interaction with the pulse emission of the 2^+ bands [31, 32, 35]. In particular, it was reported in [56] that when the gas was cooled and internal mirrors were used, new 3—3 and 3—2 stimulated emission bands were observed at wavelengths of 0.967 and 0.853 μ.

Knyazev carried out a most detailed study of the stimulated emission of the 1^+ nitrogen bands; his studies were carried out in our laboratory [52, 57]. He investigated in detail the stimulated emission spectrum and attributed about 110 lines to specific quantum transitions; he analyzed in detail the characteristics of stimulated emission and gas-discharge conditions; he calculated the principal parameters of stimulated emission. His detailed studies made it possible to establish reliably that the main mechanism was again direct electron impact: The laser levels were excited directly from the ground state of the molecule subject to the Franck—Condon principle. Knyazev reported in [57] that the peak output power was ~20 kW and the pulse energy was 4 mJ. However, his calculations indicated that this system should be capable of a saturated power output of ~1.5 kW $\cdot$ cm^{-3} $\cdot$ Torr^{-1} and that the efficiency should be about 1%. The experimentally achieved results were much poorer. The discrepancy was attributed to the insufficiently high gain and the consequent slow rate of growth of photon avalanches.

Since the principal properties of the stimulated emission of the 1^+ nitrogen system were investigated by Knyazev [57] sufficiently thoroughly, we decided to concentrate our attention on the influence of the gas temperature on the stimulated emission characteristics.

4. Pulse stimulated emission due to electronic transitions in the CO molecule was first reported by Mathias and Parker in 1963 [2] almost simultaneously with the discovery of infrared stimulated emission in the 1^+ nitrogen system, which was the first in which electronic transitions in molecules were utilized. This first paper on the stimulated emission in CO reported that a high-voltage pulse discharge through carbon monoxide produced 20 stimulated emission lines in three groups near the wavelengths 6606, 6068, and 5598 Å with a maximum peak power of 8 W, in the form of pulses of 180 nsec duration and 25 Hz repetition frequency. The observed lines were attributed to the Q branches of the 0—5, 0—4, and 0—3 bands of the electronic Ångstrom system $B^1\Sigma^+ - A_1\Pi$ of the CO molecule. Moreover, at least four other stimulated emission lines were found in the range 5198-5186 Å, which were attributed to the 0—2 band of the same system. The time characteristics of the stimulated emission and the recovery time of the laser were reported in [11-13]. It was found that the recovery time increased with the gas pressure and tube diameter and that the minimum recovery time of 100 μsec was obtained for a tube of 4 mm in diameter at a gas pressure of ~0.5 Torr. It was suggested in these papers that the population inversion was due to the excitation of the upper laser level by "fast" electrons and its quenching was the result of cascade population of the lower level by spontaneous radiation from the $C^1\Sigma^+$ state.

Further investigations of the stimulated emission spectrum were reported in [58]. The use of a pure gas flowing through the discharge gap and of internal mirrors revealed many new stimulated emission lines belonging to the Q and P branches of the known bands as well as new bands $0-1$ (4835 Å) and $0-0$ (4511 Å). The rotational structure of the $0-2$ band was reported in that paper for the first time.

The early investigators [2, 11-13, 58] achieved different durations of stimulated emission and found different optimal conditions. No systematic study of stimulated emission was attempted. This was due to the very low output power and the associated experimental difficulties in studies of this kind. The population inversion mechanism of this laser system was not subjected to a detailed analysis. General considerations were used in [8] to suggest that the direct electron impact once again was the dominant mechanism.

In view of this situation we decided to carry out a detailed experimental study of the stimulated emission of the Ångstrom bands of CO as a function of the discharge conditions. This investigation became possible only because the gas was cooled. If no cooling was applied, the output power was so low that any detailed studies would have been very difficult. Thus, a study of the influence of temperature on this system was essential especially as it was difficult to predict accurately how cooling would affect this CO laser system. Above all, the population inversion mechanism was not clear. Moreover, the asymmetry of the carbon monoxide molecule should result in effective generation of other molecules in CO discharges (particularly CO_2, C_2, O_2), so that considerable changes could be expected in the gas composition. Cooling could shift the chemical equilibrium. Special experiments were desirable to clear up the situation.

Considerable attention was paid to the time characteristics of stimulated emission because the published results were contradictory and the information on the active level lifetimes lead to the conclusion that it should be possible to achieve continuous emission, whereas only pulses were obtained experimentally. Finally, reliable data on the duration of the stimulated emission pulses were essential for the correct determination of the peak output power.

There was a special interest in the CO laser system because, among all lasers of this type, only the Ångstrom bands were located in the visible part of the spectrum extending throughout the range from red to blue. Molecular stimulated emission in the visible range should be of considerable interest especially if the output power was sufficiently high. Estimates indicated that a high power could be obtained from this laser.

List of Symbols

$k(\nu_0)$	gain at the center of Doppler-broadened line with a central frequency ν_0; $k_{max}(\nu_0)$ is the gain at the maximum of the rotational distribution, i.e., for $J = J_{max}$
λ, ν	wavelength and transition frequency
$\Delta\nu$	line-half-width at midamplitude ($\Delta\nu_D$ is the half-width due to the Doppler effect)
c, k, h	velocity of light, Boltzmann constant, and Planck constant
n, v, J	principal (electronic), vibrational, and rotational quantum numbers
Σ, Λ, Ω	projections of angular momenta of molecular electrons onto the line joining nuclei, representing the spin, orbital, and total momenta $\Delta J = J' - J''$, $\Delta\Lambda = \Lambda' - \Lambda''$, $\Delta\Omega = \Omega' - \Omega''$; the superscripts (', ", °) denote the quantities and parameters referring to the upper and lower active levels, respectively, and to the ground state of the molecule
A	probability of radiative decay (or Einstein coefficient for spontaneous radiation)
$q_{v'v''}$	Franck-Condon factor for vibrational transition $v' \to v''$

$S_{J'J''}$ Hönl — London factor (rotational line strength) for rotational transition $J' \to J''$; rotational branches P, Q, and R correspond to transitions with $\Delta J = -1, 0$, and $+1$

B rotational constant (B_e for equilibrium internuclear distance of an electronic state; B_v for v-th vibrational level)

FRR and SRR fast and slow rotational relaxation

T absolute temperature (in experimental data the temperature of the other wall of the discharge tube); T_r and T_g (T_{mol}) are the rotational and gas (molecular) temperatures

M molecular weight (M* is the mass of a molecule)

N, p density and pressure of the working gas (N_v is the population density of v-th vibrational level)

V voltage applied to storage capacitor

T_e, n_e electron temperature and density

σ electron-impact excitation cross section (σ_{max} is the value at the maximum of the excitation function)

τ duration (at midamplitude) of stimulated emission and current pulses

W, P average and peak stimulated (laser) emission power

$a = \dfrac{B_v' N_v'}{B_v'' N_v''}$ vibrational inversion coefficient and

$$a^* = a \frac{B_v'}{B_v''} \exp\left\{ \frac{hc}{kT} [B_v' \Omega' (\Omega' + 1) - B_v'' \Omega'' (\Omega'' + 1)] \right\}$$

$\gamma = \dfrac{V - V_0}{N}$ V_0 is a constant (5–7 kV)

1^+ and 2^+ first and second positive systems of nitrogen bands due to the $B^3\Pi_g \to A^3\Sigma_u^+$ and $C^3\Pi_u \to B^3\Sigma_g$

CHAPTER I

GAIN OF ELECTRONIC TRANSITIONS IN DIATOMIC MOLECULES

The gain is one of the most important characteristics of an active medium in any laser. It is particularly important for the lasers considered in the present paper. The characteristic complex level structure of molecules and the presence of a large number of transitions involved in population inversion and stimulated emission give rise to certain special features of the gain of molecular transitions. Therefore, it is desirable to give a general theoretical discussion of the gain of electronic transitions in diatomic molecules. Special stress will be made on the influence of temperature on the gain and on the distribution of the gain in the rotational structure of a band.

§ 1. General Formula for Gain of Electronic-Vibrational-Rotational Transitions

The classical expression for the gain at the center of a line (at $\nu = \nu_0$) is

$$k(\nu_0) = \sqrt{\frac{\ln 2}{\pi}} \frac{\lambda_0^3}{\Delta\nu} \frac{g'A}{4\pi} \left(\frac{N'}{g'} - \frac{N''}{g''} \right), \tag{1.1}$$

where A is the Einstein coefficient for the transition under consideration; N' and N'' are the populations of the upper and lower laser levels, and g' and g'' are the statistical weights of these levels. In the lasers considered here the line broadening is dominated, under typical conditions, by the Doppler effect. Consequently, the above formula includes the Doppler width $\Delta\nu_D$, which is the line width at midamplitude defined by the expression

$$\Delta\nu_D = \frac{2\nu_0}{c}\sqrt{\frac{2kT_{mol}}{M^*}\ln 2} = 7.158\cdot 10^{-7}\,\nu_0\sqrt{\frac{T_{mol}}{M}},\tag{1.2}$$

where M^* is the mass of a molecule and M is the molecular weight.

In an analysis of the gain of an electronic-vibrational-rotational transition $n'v'J' \to n''v''J''$ it is convenient to represent Eq. (1.1) in the form

$$k(\nu_0) = \frac{g_n}{g_n^s + g_n^a}\,\frac{hc^4}{8\pi k}\sqrt{\frac{M}{2\pi k}}\,\frac{A_{v'}q_{v'v''}}{\nu_0^3\sqrt{T_{mol}}}\,S_{J'J''}\left\{\frac{B_v'N_v'}{T_r'}\,\frac{\exp\left[(-B_v'hc/kT_r')\,J'\,(J'+1)\right]}{\exp\left[(-B_v'hc/kT_r')\,\Omega'\,(\Omega'+1)\right]} - \right.$$
$$\left. -\frac{B_v''N_v''}{T_r''}\,\frac{\exp\left[(-B_v''hc/kT_{r!}'')\,J''\,(J''+1)\right]}{\exp\left[(-B_v''hc/kT_r'')\,\Omega''\,(\Omega''+1)\right]}\right\}.\tag{1.3}$$

Several assumptions are made in the derivation of the above expression. Firstly, it is assumed that the Born−Oppenheimer approximation is satisfied so that the total wave function of a molecular energy state can be represented by the following product of the partial wave functions [59]:

$$\psi = \psi_e\psi_v\,\psi_r\tag{1.4}$$

where ψ_e is the electronic wave function, whereas ψ_v and ψ_r are the virational and rotational functions. This means that the dipole moment of the transition $\mathbf{R} = \int \psi'^*(x)\,\mathbf{R}(x)\,\psi''(x)\,dx$, which occurs in the expression for the Einstein coefficient of spontaneous radiation

$$A = \frac{64\pi^4}{3hc^3}\,\nu^3\,|\,\mathbf{R}\,|^2,\tag{1.5}$$

can be represented by the product [59]

$$|\mathbf{R}|^2 = |\,\mathbf{R}_e(r_{v'v''})\,|^2\cdot|\,\mathbf{R}_{v'v''}\,|^2\cdot|\,\mathbf{R}_{J'J''}\,|^2 = |\,\mathbf{R}_e(r)\,|^2\cdot q_{v'v''}\frac{S_{J'J''}}{(2J+1)},\tag{1.6}$$

where $S_{JJ''}$ is the rotational strength of a line (Hönl−London factor) and $q_{v'v''} = \left[\int \psi_v'\,\psi_v''\,dr\right]^2$ is the Franck−Condon factor.

Secondly, it is assumed that a molecule is a rigid symmetric top and that the coupling of the angular momenta of electrons corresponds to case (a) in Hund's rules [59], and the distribution of molecules between the vibrational sublevels is of the Boltzmann type. This means that the population of a J-th rotational sublevel can be given by the expression

$$N_J = \frac{N_v}{Z}\,\frac{g_n}{g_n^s + g_n^a}\,(2J+1)\exp\left\{-\frac{B_vhc}{kT_r}\,[J\,(J+1) - \Omega^2]\right\},\tag{1.7}$$

where g_n is the nuclear statistical weight of a symmetric (g_n^s) or antisymmetric (g_n^a) Λ state, depending on whether the population of a symmetric or antisymmetric level is considered; it is assumed that the degeneracy of the orbital momentum Λ is lifted; Z is the partition function. If $kT/B_vhc \gg 1$, which is always well satisfied, we have

$$Z \approx \frac{kT_r}{B_vhc}\exp\left\{-\frac{B_vhc}{kT_r}\,\Omega\right\}.\tag{1.8}$$

The factor in Eqs. (1.3) and (1.7), which is composed of the statistical nuclear weights, is meaningful only for symmetric molecules [59] with identical nuclei (N_2, O_2, etc.).* It governs the sequence of intensities in molecular stimulated emission spectra. This sequence depends on the nature of a transition, or more exactly, on the properties of the active levels; for states with $\Lambda = 0$, i.e., Σ states, strong and weak lines alternate with each number J (for example, all the lines with the even numbers of J are stronger and those with the odd numbers are weaker or the other way around); for states with $\Lambda \neq 0$, i.e., for Π, Δ, and other states, each rotational sublevel emits two lines (Λ doublet), one of which is weaker and the other strong. An example of the first case is the 1^+ system of nitrogen due to the $B^3\Pi_g - A^3\Sigma_u^+$ transitions. In this case the alternation of intensities is governed by the lower Σ state. Transitions of this kind readily exhibit this alternation of intensities because the separation between the strong and weak lines is simply equal to the separation between the neighboring rotational components of the band. In the second case the alternation of intensities is more difficult to detect because it is governed by the Λ doubling and the latter is so weak that the Λ components of one vibrational sublevel frequently coincide within limits of resolution of spectroscopic instruments. An example of transitions of this kind is the 2^+ system of nitrogen due to the $C^3\Pi_u - B^3\Pi_g$ transitions.

The formula (1.3) is derived on the assumption that before the arrival of an excitation pulse all the molecules are in the state of thermodynamic equilibrium described by a molecular (gas) temperature T_{mol}. Excitation makes the distribution of the molecules between the levels depart from equilibrium. However, we shall assume that the distribution between the translational degrees of freedom remains unchanged and the distribution between the rotational sublevels of different states can be described by the Boltzmann formula with a definite temperature. Consequently, Eq. (1.3) includes three temperatures: T_{mol}, T_r', and T_r''. The first of them represents the distribution of the molecules between the vibrational degrees of freedom, described by the Maxwell formula, and it occurs in the expression for the Doppler line width (1.2); the second and third describe the Boltzmann distribution of molecules between the rotational sublevels of the upper and lower active vibrational levels.

If the rotational relaxation times of these levels are considerably shorter than the level decay times (fast rotation relaxation, abbreviated to FRR), a Boltzmann distribution of the rotational sublevels with a temperature $T = T_r = T_{mol}$ is established in all the vibrational levels. In the opposite limiting case (slow rotational relaxation — SRR) the values of T_r' and T_r'' are governed by the mechanism of excitation of the active levels and can differ from one another and from T_{mol}. In the direct electron-impact excitation of a molecule from the ground state the distribution between the vibrational sublevels of the excited state is the same [59, 60] as the distribution of the populations in the ground state of the molecule. This corresponds to the conditions $T_r' = T_{mol}(B_v'/B_0^o)$ and $T_r = T_{mol}(B_v''/B_0^o)$, where B_0^o is the rotational constant of the ground state of the molecule. The laser level lifetimes and the time characteristics of stimulated emission are such that we can expect either of these two rotational relaxation cases. We shall later discuss the two limiting cases (fast and slow) of rotational relaxation separately. At this stage we shall assume that the active levels of a molecule are excited by electron impact directly from the ground state and we can assume formally that in either limiting case of rotational relaxation the distribution of the populations between all the states is described by just one temperature, which is the gas temperature $T = T_{mol}$, and the rotational constants are either the same for each state (FRR) or they should be replaced with B_0^o (SRR). Bearing

---

* In the case of asymmetric molecules, such as CO, this factor is simply omitted.

this point in mind, we can rewrite Eq. (1.3) in the form

$$k(\nu_0) = Q\sqrt{M}\ \frac{g_n}{g_n^s + g_n^a}\ \frac{A_{\ell}q_{v'r''}}{\nu_0^3 T\sqrt{T}}\ B_{\ell}'' N_{\ell}'' S_{J'J''} \left\{ a\ \frac{\exp\left[(-B_{\ell}' hc/kT)\ J'\ (J'+1)\right]}{\exp\left[(-B_{\ell}' hc/kT)\ \Omega'\ (\Omega'+1)\right]} - \frac{\exp\left[(-B_{\ell}'' hc/kT)\ J''\ (J''+1)\right]}{\exp\left[(-B_{\ell}'' hc/kT)\ \Omega''\ (\Omega''+1)\right]} \right\}, \quad (1.9)$$

where $Q = hc^4/8\pi k\ (2\pi kN^*)^{1/2} = 6.7475 \cdot 10^{25}\ cm^3 \cdot deg^{3/2} \cdot sec^{-2}$; N^* is the Avogadro number; $a = N_v' B_v'/N_v'' B_v''$ is the vibrational inversion coefficient.

§2. Distribution of Gain in Rotational Structure

of Molecular Bands

We shall now analyze the expression in the braces of Eq. (1.9), multiplied by $S_{J'J''}$. Figure 1 shows a typical dependence of the gain for a single rotational transition on the rotational quantum number J. We can see that the distribution of the gain in the band is similar to the corresponding distribution of the intensity of spontaneous radiation in a molecular band, i.e., it is bell-shaped. The rotational quantum number which corresponds to the highest gain in this band is denoted by J_{max} and the corresponding gain by $k_{max}(\nu_0)$. We must stress that the shape of the curve in Fig. 1 depends only on the quantities which occur in the product $S_{J'J''}\{...\}$ in Eq. (1.9). All the other terms in this expression are only affected in respect of the ordinate scale. The selection rules for radiative dipole transitions in molecules are $\Delta J = J' - J'' = 0, \pm 1$ and $\Delta\Omega = \Omega' - \Omega'' = 0, \pm 1$. The series of transitions for which $\Delta J = -1$ are known as the P branches, whereas those with $\Delta J = 0$ and $\Delta J = 1$ are known as the Q and R branches, respectively. Figure 2 gives the rotational distributions of the gain for the P, Q,

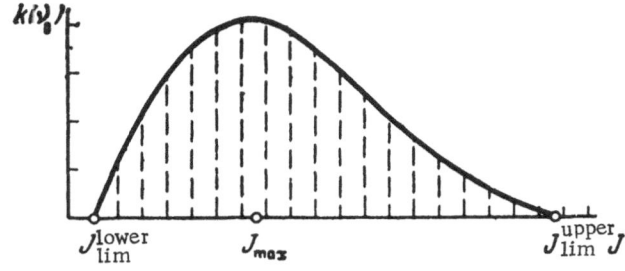

Fig. 1. Typical distribution of the gain in the rotational structure of an electronic-vibrational band.

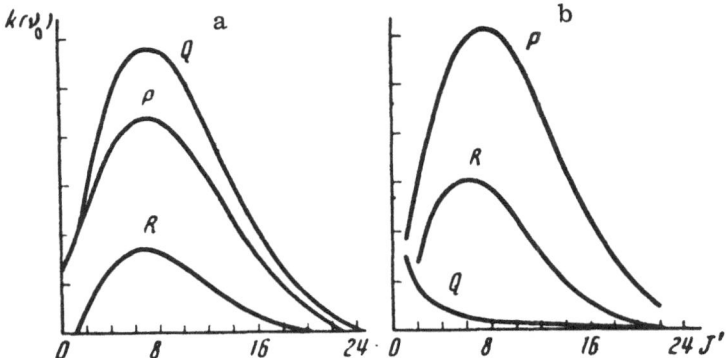

Fig. 2. Distributions of the gain in the rotational P, Q, and R branches of electronic transitions with $\Delta\Omega = -1$ and 0 at $T = 315°K$ and $a = 1.5$: a) $\Delta\Omega = -1$, CO, $B^1\Sigma^+ - A^1\Pi$ Ångstrom system, 0–4 band; b) $\Delta\Omega = 0$, N_2, $C^3\Pi_u - B^3\Pi_g$ system, 0–0 band.

TABLE 1. Hönl−London Factors $S_{J'J''}$ (Rotational Line Strengths)
for Hund's Case (a)

Branch $(J = J')$	$^{2S+1}X_\Omega \to {}^{2S+1}X_\Omega$	$^{2S+1}X_{\Omega'} \to {}^{2S+1}X_{\Omega''}$	
	$\Delta\Omega = 0$	$\Delta\Omega = +1$	$\Delta\Omega = -1$
$P(J)$	$\dfrac{(J+1)^2 - \Omega^2}{(J+1)}$	$\dfrac{(J-\Omega+1)(J-\Omega+2)}{2(J+1)}$	$\dfrac{(J+\Omega+1)(J+\Omega+2)}{2(J+1)}$
$Q(J)$	$\dfrac{\Omega^2(2J+1)}{J(J+1)}$	$\dfrac{(J-\Omega+1)(J+\Omega)(2J+1)}{2J(J+1)}$	$\dfrac{(J+\Omega+1)(J-\Omega)(2J+1)}{2J(J+1)}$
$R(J)$	$\dfrac{J^2 - \Omega^2}{J}$	$\dfrac{(J+\Omega-1)(J+\Omega)}{2J}$	$\dfrac{(J-\Omega-1)(J-\Omega)}{2J}$

Note. It is assumed in all formulas that $\Omega = \Omega'$.

and R branches of a singlet transition $\Delta\Omega = -1$ ($B^1\Sigma^+ - A^1\Pi$ transition in CO) and for a multiplet transition $\Delta\Omega = 0$ ($C^3\Pi_u - B^3\Pi_g$ transition in N_2). It is clear from the curves in Fig. 2 that the three branches are not equivalent. This is mainly due to the Hönl−London factor $S_{J'J''}$ whose values depend on the rotational branch and on the electron transition [61].

    Table 1 and Fig. 3 give analytic and graphical representations of the Hönl−London factors for the $\Delta\Omega = -1$ ($\Sigma \to \Pi$ type) transitions and for multiplet transitions with $\Delta\Omega = 0$ (where $\Omega = 0, 1, 2$). It is clear from Fig. 3a that in the case of $^3\Pi(a) - {}^3\Pi(a)$ transitions [such as those in the 2^+ nitrogen system where the upper and lower states correspond to the Hund's case (a)], the P and R branches are most likely to emit stimulated radiation, whereas Q branches can hardly be expected to exhibit population inversion. The Hönl−London factors for the P and R branches are comparable and are even somewhat higher for the R branch. However, the

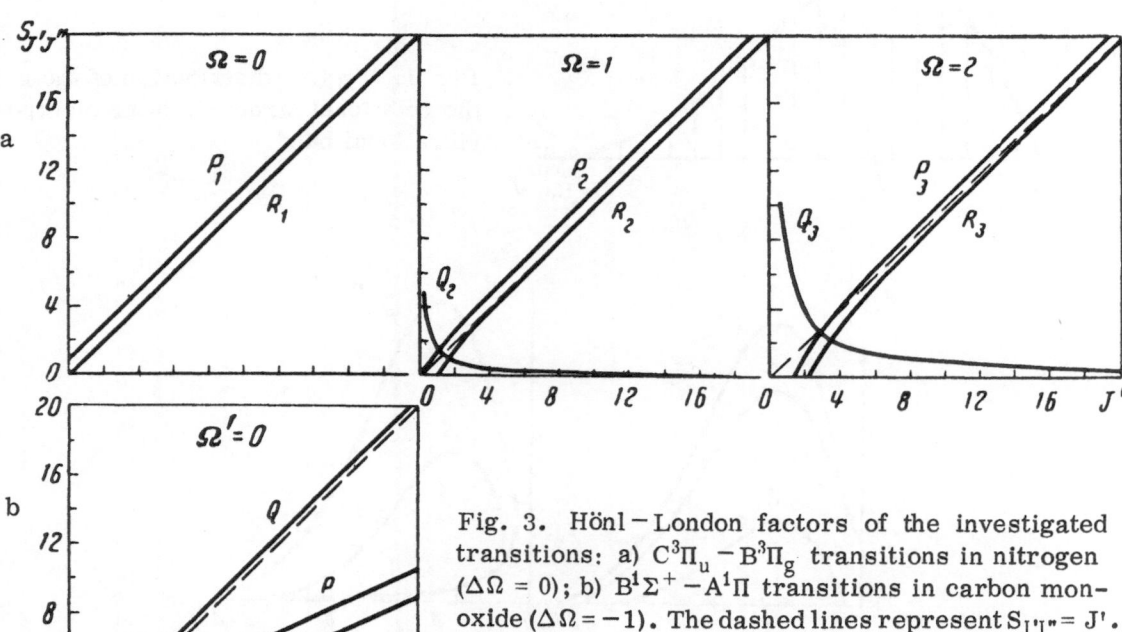

Fig. 3. Hönl−London factors of the investigated transitions: a) $C^3\Pi_u - B^3\Pi_g$ transitions in nitrogen ($\Delta\Omega = 0$); b) $B^1\Sigma^+ - A^1\Pi$ transitions in carbon monoxide ($\Delta\Omega = -1$). The dashed lines represent $S_{J'J''} = J'$.

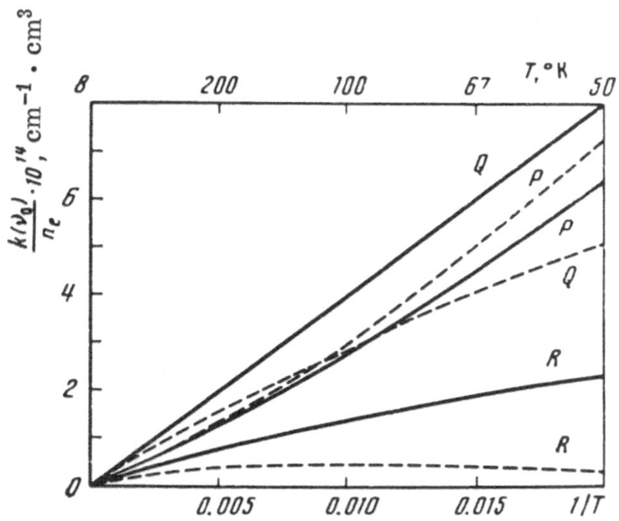

Fig. 4. Calculated dependences of the maximum gain for the P, Q, and R branches of the 0−4 band of the Ångstrom system of CO on the gas temperature. The continuous curves correspond to $a = 19$ and the dashed curves to $a = 1.5$.

gain for the R branch is less (Fig. 2b). This is due to the fact that the P branch provides more favorable conditions for population inversion because the transition takes place from a non-degenerate highly populated rotational sublevel to a sublevel with a lower population.

In the case of CO the situation is just the opposite. Here, the Q branch is the strongest, whereas the P and R branches are comparable but about twice as weak (Fig. 3b). This is typical of the $\Delta\Omega = \pm 1$ transitions. The values of $S_{J'J''}$ for $\Delta\Omega = +1$ (for example, those for $\Sigma \rightarrow \Pi$ transitions) are somewhat higher in the case of the P branch than for the R branch, whereas in the $\Delta\Omega = +1$ (for example, $\Pi \rightarrow \Sigma$) case the reverse is true. The latter applies to the first positive system of nitrogen, in which the Q branch is the strongest [52]. These relationships are always practically satisfied by the gain. However, in principle, we can have cases in which the gain of the P branch lines may nevertheless exceed the gain of the Q branch lines. This should occur as a result of strong cooling of the active gas and the higher the value of the vibrational inversion coefficient a, the stronger is the required cooling (Fig. 4).

The position of J_{max} in the rotational gain band requires separate analysis. In the spontaneous radiation spectra [59] the position of the intensity maximum in the band corresponds to the rotational transition whose quantum number is J^{sp}_{max} and whose value depends only on B'_v and T. It is either equal (Q branches) or is close (P and R branches) to the quantum number J^{NJ}_{max} of the sublevel with the highest population, which can be described quite well by [59]

$$J^{NJ}_{max} \approx \sqrt{\frac{kT}{2B'_v hc}} - \frac{1}{2}.$$
(1.10)

The value of J_{max} corresponding to the gain maximum thus depends also on the total angular momentum of electrons and on the ratio between the rotational constants and total populations of the active vibrational levels, i.e., on the parameter

$$\left(\frac{B'_v}{B''_v}\right)^2 \frac{N'_v}{N''_v} \exp\left\{\frac{hc}{kT}\left[B'_v\Omega'(\Omega'+1) - B''_v\Omega''(\Omega''+1)\right]\right\} = \bar{a}^*.$$

The analytic expressions for J'_{max} are relatively simple formulas (Table 2). These formulas correspond to the cases when $\Delta\Omega = 0$ and ± 1 and the relationship between the angular momenta of electrons in both states obeys Hund's case (a). It is assumed that either one of the states (case $\Delta\Omega = \pm 1$) or both (case $\Delta\Omega = 0$) are characterized by $\Omega = 0$. The formulas are derived on

TABLE 2

Branch $(J = J')$	General formula	Formula for $\tilde{a}^* \to \infty$
	$\Delta\Omega = \Omega' - \Omega'' = 0;\ \Omega' = \Omega'' = 0$	
$P(J)$	$J^P_{max} \approx \sqrt{\dfrac{kT}{2B_0''hc}\left(\dfrac{\bar{a}-1}{\bar{a}^*-1}\right) + \dfrac{1}{16}\left(\dfrac{\bar{a}^*+1}{\bar{a}^*-1}\right)^2} - \dfrac{1}{4}\left(\dfrac{3\bar{a}^*-5}{\bar{a}^*-1}\right)$	$J^P_{max} \approx \sqrt{\dfrac{kT}{2B_0'hc} + \dfrac{1}{16}} - \dfrac{3}{4}$
$R(J)$	$J^R_{max} \approx \sqrt{\dfrac{kT}{2B_0''hc}\left(\dfrac{\bar{a}-1}{\bar{a}^*-1}\right) + \dfrac{1}{16}\left(\dfrac{\bar{a}^*+1}{\bar{a}^*-1}\right)^2} - \dfrac{1}{4}\left(\dfrac{\bar{a}^*+1}{\bar{a}^*-1}\right)$	$J^R_{max} \approx \sqrt{\dfrac{kT}{2B_0'hc} + \dfrac{1}{16}} - \dfrac{1}{4}$
	$\Delta\Omega = -1;\ \Omega' = 0$	
$P(J)$	$J^P_{max} \approx \sqrt{\dfrac{kT}{2B_0''hc}\left(\dfrac{\bar{a}-1}{\bar{a}^*-1}\right) + \dfrac{1}{16}\left(\dfrac{3\bar{a}^*-1}{\bar{a}^*-1}\right)^2} - \dfrac{1}{4}\left(\dfrac{5\bar{a}^*-7}{\bar{a}^*-1}\right)$	$J^P_{max} \approx \sqrt{\dfrac{kT}{2B_0'hc} + \dfrac{9}{16}} - \dfrac{5}{4} \approx J^Q_{max} - \dfrac{3}{4}$
$Q(J)$	$J^Q_{max} \approx \sqrt{\dfrac{kT}{2B_0''hc}\left(\dfrac{\bar{a}-1}{\bar{a}^*-1}\right)} - \dfrac{1}{2}$	$J^Q_{max} \approx \sqrt{\dfrac{kT}{2B_0'hc}} - \dfrac{1}{2} \equiv J^Q_{max}$
$R(J)$	$J^R_{max} \approx \sqrt{\dfrac{kT}{2B_0''hc}\left(\dfrac{\bar{a}-1}{\bar{a}^*-1}\right) + \dfrac{1}{16}\left(\dfrac{3\bar{a}^*-1}{\bar{a}^*-1}\right)^2} + \dfrac{1}{4}\left(\dfrac{\bar{a}^*-3}{\bar{a}^*-1}\right)$	$J^R_{max} \approx \sqrt{\dfrac{kT}{2B_0'hc} + \dfrac{9}{16}} + \dfrac{1}{4} \approx J^Q_{max} + \dfrac{3}{4}$
	$\Delta\Omega = +1;\ \Omega'' = 0$	
$P(J)$	$J^P_{max} \approx \sqrt{\dfrac{kT}{2B_0''hc}\left(\dfrac{\bar{a}-1}{\bar{a}^*-1}\right) + \dfrac{1}{16}\left(\dfrac{\bar{a}^*-3}{\bar{a}^*-1}\right)^2} - \dfrac{1}{4}\left(\dfrac{\bar{a}^*-3}{\bar{a}^*-1}\right)$	$J^P_{max} \approx \sqrt{\dfrac{kT}{2B_0'hc} + \dfrac{1}{16}} - \dfrac{1}{4} \approx J^Q_{max} + \dfrac{1}{4}$
$Q(J)$	$J^Q_{max} \approx \sqrt{\dfrac{kT}{2B_0''hc}\left(\dfrac{\bar{a}-1}{\bar{a}^*-1}\right)} - \dfrac{1}{2}$	$J^Q_{max} \approx \sqrt{\dfrac{kT}{2B_0'hc}} - \dfrac{1}{2} \equiv J^Q_{max}$
$R(J)$	$J^R_{max} \approx \sqrt{\dfrac{kT}{2B_0''hc}\left(\dfrac{\bar{a}-1}{\bar{a}^*-1}\right) + \dfrac{1}{16}\left(\dfrac{\bar{a}^*-3}{\bar{a}^*-1}\right)^2} - \dfrac{1}{4}\left(\dfrac{3\bar{a}^*-1}{\bar{a}^*-1}\right)$	$J^R_{max} \approx \sqrt{\dfrac{kT}{2B_0'hc} + \dfrac{1}{16}} - \dfrac{3}{4} \approx J^Q_{max} - \dfrac{1}{4}$

Note. Here $\bar{a}^* = \bar{a}\dfrac{B_0'}{B_0''} = a\dfrac{B_0'}{B_0''}\exp\left\{\dfrac{hc}{kT}\left[B_0'\Omega'(\Omega'+1) - B_0''\Omega''(\Omega''+1)\right]\right\}$, where $a = \dfrac{B_0'N_0'}{B_0''N_0''}$.

the assumption that the expression exp $[\Delta F(J)/kT]$ varies slowly in the region of J_{max} and remains ≈ 1 $[\Delta F(J) = F'(J') - F''(J'')$ is the difference between the energies of the rotational sublevels participating in the transition under consideration]. This approximation makes it possible to use the formulas of Table 2 in finding fairly accurate values of J_{max}, particularly for large values of $a^*(\gtrsim 3)$, and in the limiting case $a^* \to \infty$ they transform simply into expressions for J_{max}^{sp} of the corresponding spontaneous transitions. A specific calculation for the $0-0$ band of the $C^3\Pi_{u0} - B^3\Pi_{g0}$ system of nitrogen and of the $0-4$ band of the $B^1\Sigma_0^+ - A^1\Pi_1$ system of carbon monoxide at $T = 315°K$ carried out using the formulas of Table 2 for the first five branches gives the results shown graphically in Fig. 5. The thick curves in Fig. 5 correspond to the FRR case and the thin ones to SRR, when we can arbitrarily assume that $B_v' = B_v'' = B_0^\circ$. The dotted lines represent the values of J_{max}^{sp} whereas the dashed lines are the corresponding curves found by graphical determination of J_{max} from series of rotational gain bands calculated by applying Eq. (1.9) to various values of a. It is clear from Fig. 5 that for $\tilde{a}^*$ ranging from 1 to ~3 the value of J_{max} depends strongly on N_v'/N_v'' and it may reach very high or very low values at the same temperature and for the same rotational constants. The value of J_{max} remains the same for different vibrational inversions only for the Q branch in the FRR case. It is then identical with J_{max}^{NJ} [see Eq. (1.10)].

It is also worth noting an additional difference between distributions of the rotational spontaneous line intensities and the gain. In spontaneous radiation a given band represents, in principle, an infinite number of rotational sublevels because their population approaches zero asymptotically when J increases. In the stimulated emission case only a limited number of rotational transitions participates in population inversion and this number depends on the rotational constants and the gas temperature, as well as on the vibrational inversion coefficient. Figure 6 shows several dependences of J_{lim}, which are the limiting values of the rotational

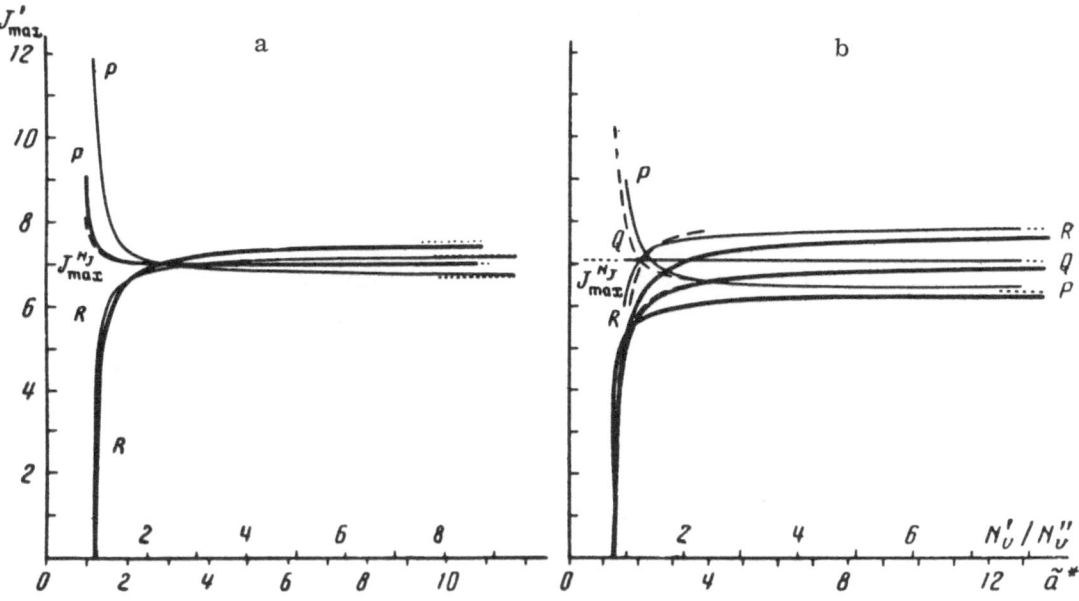

Fig. 5. Dependences of the rotational quantum number J'_{max} on the vibrational inversion ($T = 315°K$): a) $\Delta\Omega = 0$, P_1 and R_1 branches of the $0-0$ band of the 2^+ system of N_2; b) $\Delta\Omega = 1$, P, Q, and R branches of the $0-4$ band of the Ångstrom system of CO. The continuous curves represent the results of calculations carried out using approximate formulas of Table 2; the thick curves correspond to the FRR case and the thin ones to SRR; the dashed lines give the exact values calculated using Eq. (1.9).

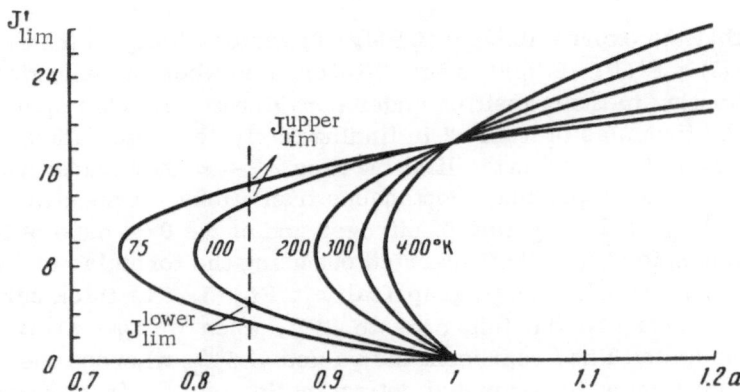

Fig. 6. Dependences of the limiting values of the rotational quantum number, between which population inversion is observed, on the vibrational inversion coefficient, plotted for various gas temperatures (P_1 branch of the 0−0 band in the 2^+ system of nitrogen, FRR case).

quantum number between which population inversion occurs, on the value of a at different gas temperatures; the results are plotted for the P_1 branch of the 0−0 band in the 2^+ system of nitrogen under FRR conditions. It is worth noting two features. Firstly, near $a \approx 1$ the upper value of J_{lim} is independent of T and it is equal to ~18 (for the lower level). Secondly, at any given temperature only one line $J_{lim}^{''} = J_{cr}^{''} \approx 9$ exhibits population inversion when a has its minimum value a_{min}. In the SRR case the dependence of J_{lim} on a is somewhat different: In particular, we find that $J_{cr} \rightarrow \infty$ in the limit $a \rightarrow a_{min}$. We can show that for $\Delta\Omega = 0$ and the P branches

$$J_{cr}^{''} = \frac{B_v' + B_v''}{2\,(B_v' - B_v'')}.$$ (1.11)

It is clear from this formula that J_{cr} tends to infinity in the SRR case.

§ 3. Dependence of Gain on Gas Temperature

It is clear from Eq. (1.9) that the expression for the gain includes an explicit temperature dependence. Moreover, some of the quantities in this equation may depend implicitly on temperature. The occurrence of population inversion and the value of the gain are governed by the difference which is in the braces, and they depend strongly on the parameter a. This parameter is governed primarily by the ratio N_v'/V_v''. The populations of the vibrational levels N_v' and N_v'' are themselves governed by the excitation and decay conditions. If, for example, the excitation of all the levels under discussion proceeds directly from the ground state of the molecule and it is due to electron impact and if we postulate that the population of the ground state remains constant and the decay (deactivation) of the levels results solely from the emission of spontaneous radiation, we can assume that N_v' and N_v'' are independent of the gas temperature. However, if collisions of molecules with other molecules (or with heavy particles) play an important role in the excitation and decay of the active levels, the populations N_v' and N_v'' are generally dependent on T because the collision cross sections depend on the velocity of the colliding particles.

The nature of this dependence is governed by the mechanisms of the populations and decay (deactivation) of the active levels and cannot be described by a general expression. Thus, the dependence of the gain (and other laser characteristics) on the laser temperature varies with the population inversion mechanism.

We shall assume that the active levels are excited by direct electron impact from the ground state and that they decay by emitting spontaneous radiation. Consequently, we shall assume that $N_v^!$ and $N_v^"$ are independent of temperature. A comparison of the results of such calculations with the experimental data reveals in which cases this assumption is justified, so that the population inversion mechanism can be identified.

It follows from Eq. (1.9) that the temperature dependences of the gain should vary from line to line. A specific calculation of the distribution of the gain in a band at various temperatures is given later. However, it is interesting to consider the temperature dependence of the maximum value of the gain, i.e., of J_{max}. It follows from the above formulas that in the cases of interest to us we can quite accurately assume that the Hönl–London factors increase linearly with J'. If we adopt this approximation and use the formula for J_{max} from Table 2, we can show by substitution in Eq. (1.9) that in the case of high vibrational inversion coefficients ($a \gg 1$) the maximum gain of all the branches can be approximated satisfactorily by

$$k_{max}(\nu_0) \simeq \frac{a \Delta N_r}{T}. \qquad (1.12)$$

Thus, at high values of a the gain at the maximum of the rotational distribution is approximately inversely proportional to the gas temperature. This is due to the influence of two factors, one of which is the temperature dependence of the Doppler line width and the other is the redistribution of the population between the rotational sublevels of the active vibrational levels. These two factors make approximately the same contributions.

If a is close to unity, we have to include also the second term in the braces of Eq. (1.9). In this case the temperature dependence of the gain becomes very complex and specific calculations are needed in each case. An example of such a calculation is given below for the 2^+ nitrogen system.

§ 4. Calculation of Gain for 0 − 0 Band in Second
Positive System of Nitrogen

We shall now describe the specific calculation of the gain for the P branches of the strongest (0−0) ultraviolet stimulated emission band in the 2^+ nitrogen system ($C^3\Pi_u - B^3\Pi_g$ transition). The calculation is made on the assumption that the active levels are populated by electron impact from the ground state of the molecule $X^1\Sigma_g^+ (v = 0)$ and that consequently the ratio of the total populations of the upper and lower active vibrational levels $N_v^!/N_v^"$ is independent of the gas temperature. Only the P branches are considered because for the selected electronic transition the gain is highest for these branches (see Fig. 2b and Chap. III). The calculation is made for two extreme cases of the fast and slow rotational relaxation. These two cases are of definite practical interest. In the case of a considerable difference between the spectral characteristics in these two cases we can use the stimulated emission spectra in estimating the rotational relaxation rates of excited electronic states of molecules.

An analysis of the temperature dependence of the gain can conveniently be started from the case when $N_v^! \gg N_v^"$, i.e., when the vibrational inversion coefficient is $a = N_v^! B_v^! / N_v^" B_v^" \gg 1$. Then, the second term in the braces of Eq. (1.9) can be dropped. The dependence of the gain on J' is similar to the dependence of the intensity of this spontaneous radiation starting from the upper level. For a P_1 branch (transition characterized by $J' \to J" + 1$ and $\Omega' = 0 \to \Omega" = 0$) in the FRR case, we have

$$k(\nu_0) = Q_{v'v"} \frac{B_e' N_e'}{T^{3/2}} (J' + 1) \exp\left\{ -\frac{B_e' hc}{kT} J'(J' + 1) \right\}, \qquad (1.13)$$

where

$$Q_{v'v''} = Q\sqrt{M}\,\frac{g_\mathrm{n}}{g_\mathrm{n}^s + g_\mathrm{n}^a}\,\frac{A_{v'}q_{v'v''}}{\nu_0^3}.$$

It is clear from Table 2 that the rotational number $J'_{\max}$ corresponding to the maximum gain is approximately given by the expression

$$J'_{\max} = \sqrt{\frac{kT}{2B'_v hc}} - \frac{3}{4}. \tag{1.14}$$

In the SRR case we have to replace B'_v with B°_0, which does not result (for the transition under consideration) in a significant change in $J'_{\max}$ because $B'_v = 1.826$ cm^{-1} and $B^\circ_0 = 2.010$ cm^{-1} [59]. If Eq. (1.13) is obeyed, it follows that for $J_{\max}$ and $kT/Bhc \gg 1$ (this condition is usually well satisfied) we have $k_{\max}(\nu_0) \propto T^{-1}$, i.e., at high values of the vibrational inversion coefficient the maximum gain is inversely proportional to the absolute temperature. The maximum gain in the FRR and SRR cases differ only slightly if $a \gg 1$.

If the vibrational inversion coefficient is low, the second term in Eq. (1.9) cannot be ignored and it is difficult to obtain an analytic dependence of $k_{\max}(\nu_0)$ on T. Therefore, we carried out direct numerical calculations of the maximum gain at various temperatures. The results of this calculation, carried out using Eq. (1.9) for the two rotational relaxation cases and $a = 1$, are plotted in Fig. 7. We can see that the situation is still qualitatively the same: the maximum gain increases when the gas temperature is lowered. However, the rate of this increase is now different. In the temperature range from 400 to 50°K we find that $k_{\max}(\nu_0) \propto T^{-1.75}$, which is true to within 5%. There is little difference between the FRR and SRR cases. The gain is slightly higher for the SRR case but the rate of rise is the same in both cases.

Thus, such calculations demonstrate that when the vibrational inversion coefficient is low, the maximum gain rises with decreasing gas temperature twice as fast as in the strong vibra-

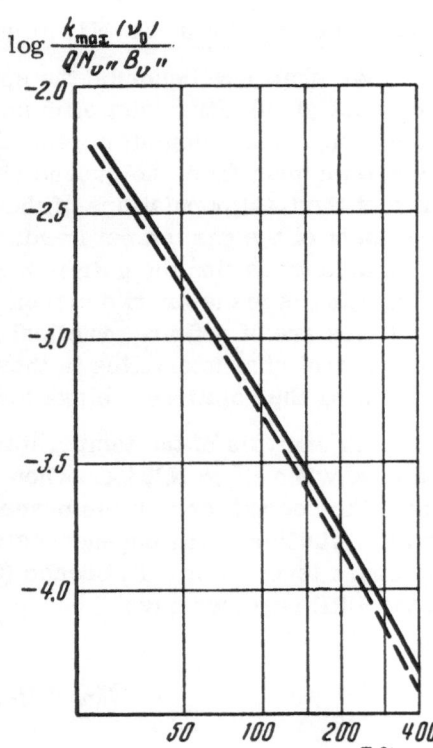

Fig. 7. Dependences of the maximum gain on the gas temperature. The continuous curve corresponds to slow rotational relaxation and the dashed curve to fast relaxation. The dependences are plotted for the 0–0 band in the 2^+ system of nitrogen, $a = 1$.

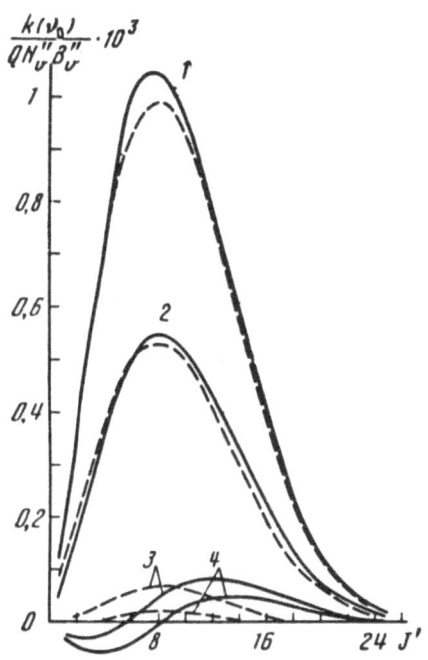

Fig. 8. Dependences of the gain on the upper rotational quantum number at T = 300°K, plotted for different values of the vibrational inversion coefficient a: 1) 2; 2) 1.5; 3) 1; 4) 0.95. The continuous curves apply to slow rotational relaxation and the dashed curves to fast relaxation (0−0 band in the 2^+ system of N_2).

tional inversion case. For example, cooling from 320 to 80°K increases $k_{max}(\nu_0)$ by a factor of 13 if $a = 0$ and by a factor of 4 if $a \gg 1$.

The greatest difference between the SRR and FRR cases is observed in the rotational structure of the gain at low values of the vibrational inversion coefficient ($a \approx 1$). It is clear from Fig. 8, which shows the rotational structure of the gain for different values of a at T = 300°K, that the differences between the extreme rotational relaxation cases are negligible for $a \geq 1.5$. When a increases still further, the values of J'_{max} become practically identical in both cases and the gain at the maximum of the distribution in the SRR case remains approxi-

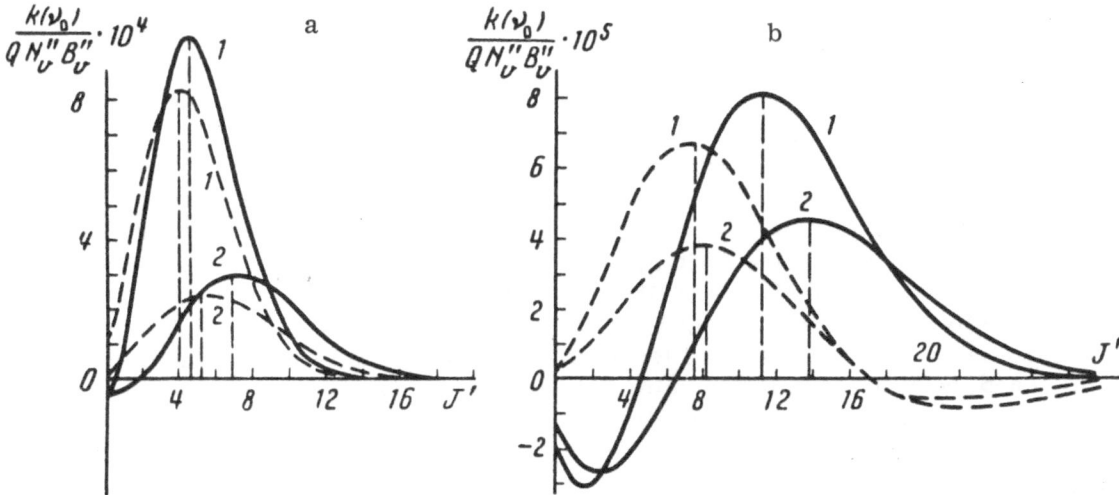

Fig. 9. Dependences of the gain on the rotational quantum number of the upper level at different gas temperatures (°K): a1) 75; a2) 150; b1) 300; b2) 400. The continuous curves represent slow rotational relaxation and the dashed curves fast relaxation (0−0 band in the 2^+ system of N_2, $a = 1$).

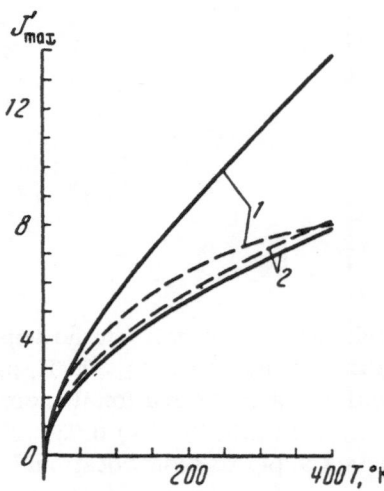

Fig. 10. Temperature dependences of the rotational quantum number J'_{max} of the upper active state of the $0-0$ band in the 2^+ system of nitrogen, $a = 1$ (curves denoted by 1), and of the same quantum number for spontaneous radiation (curves denoted by 2). The continuous curves correspond to slow rotational relaxation and the dashed curves to fast relaxation.

mately 10% higher than for FRR. However, if a is close to unity, the rotational structure of the gain differs in respect of the position and its nature for fast and slow rotational relaxation. These differences are particularly marked for high temperatures and low values of J.

Figure 9 shows the calculated dependences of the gain on J' for the P_1 branch at various temperatures. These dependences are calculated for $a = 1$. We can see that when the temperature is lowered, a strong rise of the maximum gain is accompanied by a shift in the direction of lower values of J. Then J'_{max} differs quite considerably for the two extreme cases of rotational relaxation, particularly at high temperatures. The temperature dependence of J'_{max} is plotted in Fig. 10 for $a = 1$ (curves denoted by 1). For comparison, this figure includes (curves denoted by 2) the dependences for spontaneous emission from the upper level. The differences between J'_{max} are so large (for example, at room temperature $\Delta J_{max} \approx 4$) that if no forced cooling is used under conditions corresponding to $a = 1$, we can determine which rotational relaxation case applies. Moreover, the gain spectra in the SRR case are characterized by absorption at low values of J. An interesting feature of the transition under discussion is that in the FRR case for values of a tending to the minimum a_{min} we find that, irrespective of temperature, $J'_{max} \to 8$.

It is clear from Fig. 8 that, as in the case of vibrational transitions [62, 63], we find that the gain for electronic transitions is positive in a certain range of J for definite branches even if the ratio of the populations is $N'_v/N''_v < 1$ (partial inversion). For the transition under consideration we have $a_{min} = 0.93$ at 320°K and 0.75 at 80°K. For vibrational transitions within one electronic state the rotational constants of the lower and upper levels are practically equal and consequently the ratio N'_v/N''_v is practically identical with a. However, for electronic transitions the values of B'_v and B''_v may differ considerably, particularly if the potential curves of the active states are displaced significantly. Consequently, the minimum ratio of the populations of the active vibrational levels in electronic transitions may be even smaller than a_{min} and, consequently, it may differ considerably from unity. For example, for the transition in question at 80°K we find that this ratio can be ~0.61 in the SRR case, whereas $a_{min} = 0.75$.

The dependences described above are obtained for the P_1 branch, i.e., for transitions between electronic components with a total angular momentum of the upper and lower states $\Omega = 0$. It is known from the experimental results that stimulated emission is observed also for the P_2 and P_3 branches ($\Omega = 1$ and 2, respectively). A calculation carried out on the assumption that the populations of the multiplet components are equal shows that the rotational structures of all the branches are similar. This similarity is demonstrated in Fig. 11, which gives

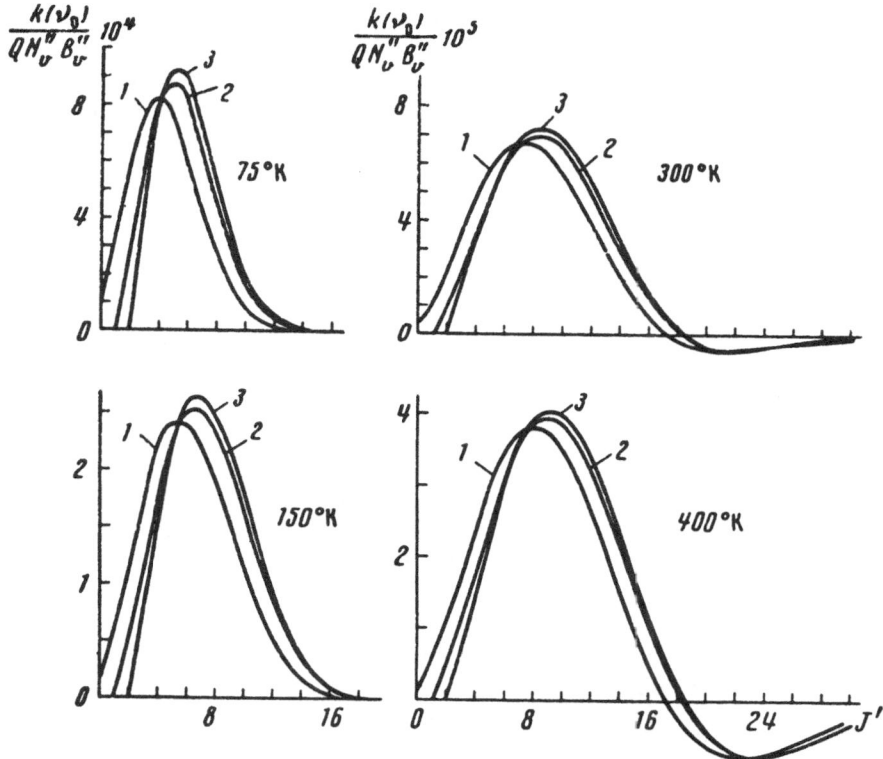

Fig. 11. Dependences of the gain for different components of a multiplet on the rotational quantum number J' at four gas temperatures. Curves 1, 2, and 3 apply to the P_1 ($\Omega = 0$), P_2 ($\Omega = 1$), and P_3 ($\Omega = 2$) branches of the 0—0 band in the 2^+ system of nitrogen (fast rotational relaxation).

the relevant curves for $a = 1$ at four different temperatures. The strongest branch is P_3 and the weakest one is P_1. The position of J_{max} is practically the same for P_2 and P_3 branches but in the case of the P_1 branch it is shifted slightly in the direction of lower values of J. This difference between the branches is not greatly affected by the gas temperature. As pointed out earlier, the calculations are made on the assumption of equality of the vibrational populations of the various electronic multiplet components differing in respect of the total angular momentum Ω. However, if an equilibrium is established between these components, the ratio of the vibrational populations should be governed by the statistical weights $2\Omega + 1$. However, in the case when this equilibrium is not established during a stimulated emission pulse, the ratio depends on the population inversion mechanism and on the selection rules.

We can summarize the results of this chapter by saying that if the active electronic states of molecules are excited directly by electrons from the ground state, cooling of the working gas increases considerably the gain. If the vibrational inversion coefficient is large, the maximum gain in the rotational distribution is inversely proportional to the absolute temperature, whereas for transitions characterized by small vibrational inversion coefficients, the rise of the gain as a result of cooling can be much faster. Thus, cooling of the active gas provides means for a considerable increase in the gain of lasers utilizing electronic transitions in molecules, particularly in those cases when the inversion is weak. This opens up additional opportunities in the search for new transitions and new active media. On the other hand, it is clear that special measures are needed to prevent excessive heating of the working gas in lasers of this kind.

A calculation of the gain for the $0-0$ ultraviolet band in the second positive system of nitrogen shows that the distribution of the gain in a rotational band and the maximum gain as well as its rise as a result of cooling of the gas are similar for the two extreme (fast and slow) cases of rotational relaxation. For example, if $a > 1.5$, the differences do not exceed 10%. Considerable differences (in the rotational structure of the stimulated emission spectrum) should be manifested if the vibrational inversion coefficient is very small and the temperature is high. Then, the extreme cases of the rotational relaxation differ so much that, in principle, it should be possible to detect this difference experimentally and thus estimate the rate of rotational relaxation of the excited molecules.

CHAPTER II

EXPERIMENTAL METHOD

§ 1. Discharge Excitation System

One of the advantages of pulse operation of a laser is the ability to use very high instantaneous excitation powers. This can be done if a suitable high-power pulse supply system is available. Since our aim was not to obtain the best possible results for any one laser but to carry out a wide range of studies on many laser systems, including new laser emission lines, the main requirements that our pulse supply system had to satisfy were a sufficiently high pulse power, universality of use, simplicity, and convenience.

The system necessarily had to be a high-voltage one because it had to supply high instantaneous excitation powers. We used a pulse generator based on the discharge of a high-voltage capacitor through a tube containing the working gas. A three-electrode air discharger was a switching element. This system was characterized by a relative simplicity and the ability to generate very high pulse currents. Its main disadvantages were relatively large size and a relatively low maximum pulse repetition frequency.

Schematic diagrams of the whole system and of the power supply are shown in Fig. 12. A high-voltage rectifier HVR charged, through a charging resistor R_{ch}, a working capacitor C_w to a voltage which could be as high as 55–60 kV. A buffer capacitor C_b was used to ensure a constant voltage at the HVR output and to prevent large instantaneous overloads of kenotrons used in the rectifier. The working capacitor C_w was discharged periodically through the tube via a simple three-electrode air discharger. Pulses which triggered off the discharger were applied to an auxiliary third electrode placed inside a recess drilled along the axis of one of the main electrodes. The discharge current was measured with a low-inductance shunt R_{sh}.

The high-voltage rectifier was based on an x-ray unit (AII-70) with an average output power of 300 W. The working capacitor was usually of the IM-110-0.01 type with a capacitance of 0.01 μF and a self-inductance of about 1 μH. Sometimes we used special low-inductance capacitors (KMP) made at the Leningrad Polytechnic Institute. Their self-inductance was about 30 nH. The charging resistor was a chain of glass-encapsulated elements amounting in total to 0.5–1.0 MΩ. The air discharger was operated at atmospheric pressure. Its main electrodes were solid brass rods with rounded ends. The distance between the electrodes was varied depending on the working voltage across the capacitor. When this distance was fixed, the discharger operated stably in a limited range of voltages (for example, in a range ± 7 kV when the average voltage was 34 kV). Triggering pulses were generated in a thyratron circuit. This circuit terminated with a step-up coreless transformer, so that voltages up to 20–30 kV could be applied to the triggering electrode. The triggering circuit was controlled by a G5-7A (GIS-2M) pulse generator. The repetition frequency was limited by the discharger

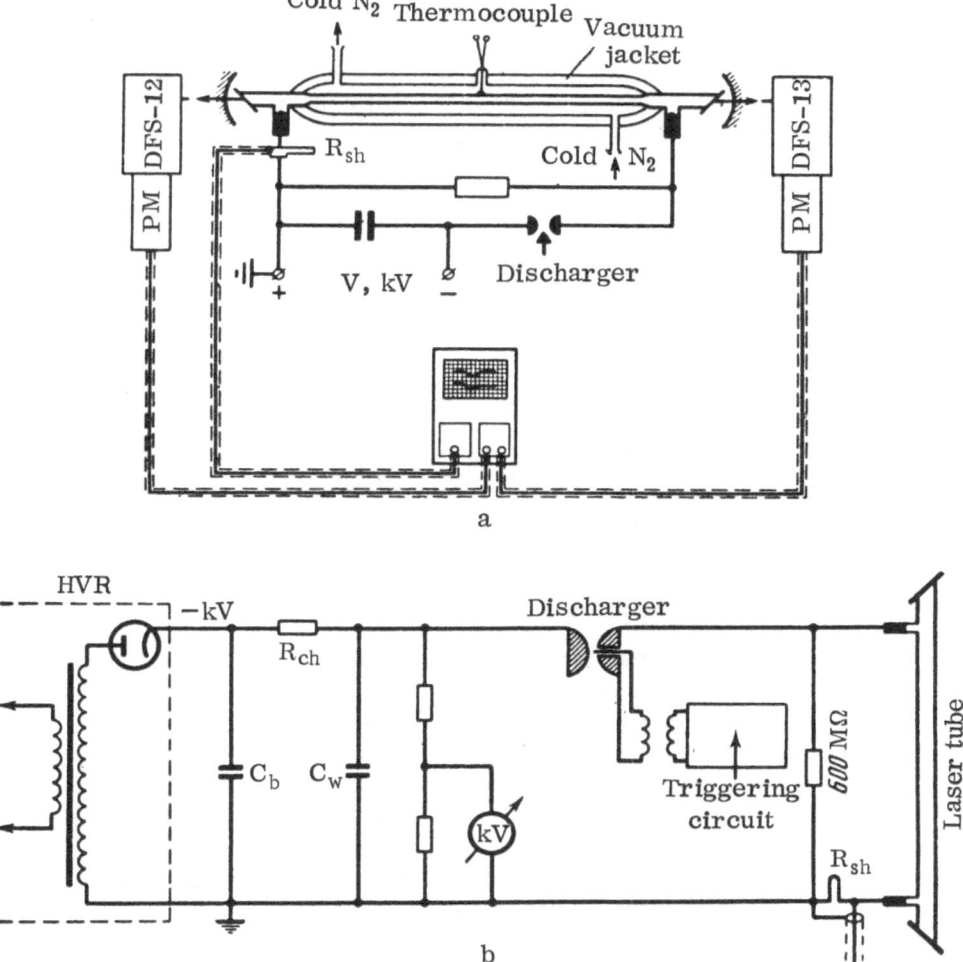

Fig. 12. Schematic diagram of the apparatus: a) whole system and construction of laser discharge tube; b) system supplying power to pulse discharge; C_b is a buffer (filtering) capacitor and C_w is the working capacitor; PM are photomultipliers; DFS-12 and DFS-13 are spectrographs; HVR is a high-voltage rectifier.

and the charging capacitor. Experience showed that our simple three-electrode discharger operated stably up to several tens of hertz. Usually the repetition frequency was lower and often 2 Hz. This low frequency was chosen to avoid overheating the gas in the tube. We found that such overheating (this point is discussed later) was particularly undesirable in the case of molecular lasers. The duration and amplitude of the current pulses depended on the nature and pressure of the working gas and also on the capacitance of C_w and the voltage applied to it. Usually the current pulses were bell-shaped with a single maximum. Current pulses in the form of rapidly damped sinusoids were also obtained when the capacitance and voltage were high. The duration of the current pulses was usually ~ 1 μsec at the base and the amplitude varied from hundreds to thousands of amperes. When the capacitance was 0.01 μF, the energy stored in the capacitor was approximately 10 J. The peak power supplied to the discharge by this system was of the order of several tens of megawatts.

Experience showed that the system employed made it possible to generate easily stimulated radiation as a result of a very large number of transitions in various gases. In particular, we were able to detect and investigate laser emission due to electronic transitions in
molecules in all known cases (apart from transitions in the vacuum ultraviolet for which stimulated emission was achieved recently [64-66] employing special power supply systems). We
also readily observed many stimulated transitions in atoms and ions in a wide spectral range
from infrared to ultraviolet.

The simplicity and universality of the power supply system just described made it convenient to use.

§ 2. Measurement of Parameters of Excitation Pulses

The most important characteristics of excitation pulses were their amplitude, shape, and
duration. In our case a discharge of a capacitor through a controlled air discharger produced
pulses whose duration was usually about 1 μsec. No special difficulties were encountered in
the measurement of current pulses of this duration. This was done mainly with a low-inductance shunt connected in series with the discharge circuit between the laser tube and the
grounded electrode (Fig. 12); we also used a Rogowski loop which operated as a current transformer. The shunt resistance was selected depending on the amplitude of the current and it
amounted to 0.02-0.2 Ω. The amplitude could be determined from the known charge stored in
the working capacitor and the relative change in the current with time.

The voltage across the capacitor C_w was measured with an S-96 static kilovoltmeter.
The voltage across the discharge tube was sometimes measured using capacitative voltage
dividers. The voltage pulses could also be reconstructed from the time dependences of the
current and, consequently, of the charge in the capacitor. In the investigated cases the stimulated radiation pulses usually appeared during the leading edge of a current pulse. The
duration of the stimulated radiation pulses emitted by most of the investigated lasers was
considerably shorter than the duration of the current pulses or even their leading edges. Therefore, in the first approximation, we could assume that the voltage across the capacitor did not
change greatly during a stimulated radiation pulse.

A detailed investigation of the processes occurring during pulse discharges was made
using apparatus designed by Knyazev [57] who was also a member of our team. He showed that
fairly complex breakdown and field establishment processes occurred in a tube under conditions
typical of our experiments. However, stimulated emission appeared during the leading edge
of a current pulse, corresponding to the high-current phase of the discharge, when the breakdown phenomena were completed and an approximately homogeneous field was established along
the tube. Thus, the active plasma in the discharge could be regarded quite accurately as a
homogeneous medium and we could consider processes per unit volume of this medium. This
simplified considerably the analysis of the population inversion processes. Knyazev [57] also
measured the electron density in the active media under conditions typical of our experiments.
Therefore, we used his results in an analysis of the excitation mechanisms in plasmas of
high-current pulse discharges in gases.

§ 3. Spectral Measurements. Spectroscopic Apparatus

Measurements of the spectral characteristics of the spontaneous and stimulated radiation played an important role in our study. Essentially, we investigated physical processes
in the active medium by analyzing the stimulated and spontaneous radiation spectra and their
dependences on the experimental conditions and on the nature of the active substance. In
identifying the population inversion mechanism it was very important to know the time characteristics of the output radiation, i.e., the position on the time axis, shape, and duration of the

light pulses representing individual spectral lines. In the present section we shall describe only the spectroscopic apparatus; the results of the determination of the time characteristics and a description of photodetectors will be given later.

Before investigating the population inversion processes in the selected transitions we had to identify these transitions reliably. This demanded a high precision in the wavelength measurement method. This was particularly difficult because the molecular spectra consisted of a large number of closely spaced lines.

Different spectroscopic apparatus was used at various stages in our investigation. Wavelengths of the newly discovered laser emission lines were measured accurately and the spectra of the known lines were identified using a DFS-13 spectrograph with a plane diffraction grating and mirror optics, designed for the range from 2000 to 10,000 Å. The focal length of the mirrors was 4 m and the relative aperture was 1:40. This spectrograph had two interchangeable gratings with 600 and 1200 lines/mm. The first-order linear dispersion of the grating with 600 lines/mm was 4 Å/mm and that of the grating with 1200 lines/mm was 2 Å/mm. The stimulated emission spectra were investigated in different diffraction orders. The highest linear dispersion was used in the investigation of ultraviolet lines emitted due to transitions in molecular nitrogen. In this case we used the third-order diffraction and the 1200 lines/mm grating, in which case the linear dispersion reached ~0.5 Å/mm. The DFS-13 spectrograph allowed us to identify reliably the stimulated emission lines of practically all the investigated media. Only in some cases were the lines belonging to different rotational branches located so close that it was impossible to separate them with this spectrograph. The wavelengths were measured usually by a photographic method in which we compared the measured spectrum with a standard one. The standard spectrum was usually the arc spectrum of iron. In some cases we used also the arc spectra of copper, zinc, and titanium. These spectra were photographed on films or plates. Photographic plates were used in the exact determination of the wavelengths. The error in such determinations was approximately ±0.03 Å, which was sufficiently small for most cases. The precision of the determination of the relative positions of the lines carried out with an IZA-2 linear comparator was an order of magnitude higher. In this case we identified the stimulated emission lines not only by absolute measurements of the wavelengths but also by direct comparisons of the stimulated and spontaneous radiation spectra. This comparison was particularly important in the case of molecular spectra with complex rotational structures. The correctness of the identification of the rotational lines of the investigated molecules was checked also by plotting the Fortrat curves.

The relative intensities of the stimulated emission lines were determined by photographing the spectra on aerial film via a nine-stage attenuator. The uniformity of illumination along the height of the entry slit of the spectrograph was ensured by focusing light onto the slit with a cylindrical lens. This uniformity was checked with an MIR-2 microphotometer, which was used also in the measurements of the relative intensities of the lines. The necessary density in the photographic emulsion was achieved and the errors due to variation from pulse to pulse were avoided by superimposing between 100 and 360 light pulses in one frame.

The emission spectra, particularly in the study of the time characteristics, were investigated using a variety of spectroscopic instruments and photoelectric recording. We employed a DFS-12 monochromator with two replica gratings (600 lines/mm) and a relative aperture of 1:5.3. The linear dispersion of this monochromator was 5 Å/mm in the working order. The monochromator allowed us to separate reliable single rotational components of molecular bands of most of the investigated spectra.

An IKS-12 prism spectrometer was used in the near infrared ($\lambda \approx 5\,\mu$). The relative aperture was 1:5.5. We used also a modified form of the DFS-13 spectrograph.

§ 4. Determination of Time Characteristics

of Spontaneous and Stimulated Radiation

Determination of the time characteristics of the output radiation is one of the more important aspects of investigations of physical processes in pulse gas lasers. For this purpose we used various photoelectric detectors at the outputs of spectroscopic instruments. The main measurements were carried out in the ultraviolet and near infrared range between 2000 and 10,500 Å. The detectors were mainly photomultipliers, denoted by PM in Fig. 12: FÉU-28 and FÉU-36 in the infrared and visible range, and FÉU-39 in the ultraviolet range. The time resolution of these photomultipliers used in combination with an S1-11 oscillograph was ~10 nsec. This estimate was deduced from a study of the 6143 Å superradiance line of neon which was emitted in the form of 10-nsec pulses.

A double-beam oscillograph of the S1-17 type was used in simultaneous measurements of the current and radiation pulses. In this case the time resolution was limited to the pass band of the oscillograph amplifier and it amounted to 40 nsec.

The duration of the spontaneous and stimulated radiation pulses was much greater than the estimated time resolution of the apparatus so that we assumed that the apparatus did not distort significantly the time characteristics of the radiation. The linearity of the time and amplitude oscillograph scales was specially checked. Calibration graphs were used in cases of departure from linearity.

In some cases other photodetectors and recording apparatus were used.

One of the main difficulties encountered in the experimental studies of pulse gas-discharge lasers excited by the application of high voltages arose from the fact that the discharge itself was a source of strong electrical strays and interference. The strays were particularly harmful when relatively weak signals were recorded using wide-band instruments. The greatest difficulties were experienced in the photomultiplier recording when the signal was applied to the input of the oscillograph amplifier (for example, when the FÉU-36 photomultiplier was used in conjunction with the S1-17 oscillograph). Strays were avoided by careful screening of all the instruments. In particularly difficult cases we screened all the connecting coaxial cables and the power supply systems. The photomultipliers were carefully enclosed in thick-walled metal tubes. Special filters were used to suppress strays which could be picked up from the line supply. Special attention was paid to the quality of all the contacts in the signal and screen circuits. The whole system was grounded at one point, which was selected so that the stray signals were as weak as possible. These measures enabled us to observe, on the oscillograph screen, undistorted time characteristics, which were particularly important for the correct understanding of the underlying physical processes.

Another important experimental task was the determination of the energy and peak power of the stimulated radiation pulses. In most cases we determined the average power at some fixed repetition frequency. The average power measurements were normally carried out employing a slow-response calibrated thermopile. The error in the absolute measurements by this method was ±20%. The light beam intensity was reduced, if necessary, by filters calibrated employing the investigated laser radiation. The energy of the stimulated radiation pulses was deduced from the average power and repetition frequency. The peak power was calculated from the pulse energy and duration.

§ 5. Gas-Discharge Tubes. Vacuum System

Quartz and molybdenum-glass discharge tubes of different lengths and diameters were used in investigating laser systems. Usually, the discharge part of relatively wide tubes (with internal diameters from 3 to 30 mm) was 1–1.2 m and never exceeded 2 m. In most cases external mirrors were used. Therefore, the discharge tube ends were oriented at the Brewster angle and were fitted with optical-quality windows, usually made of fused quartz (vitreous silica). The Brewster windows were used also in high-gain systems (for example, those used in an investigation of ultraviolet stimulated emission from nitrogen). In the latter case they were needed mainly to avoid undesirable reflections from the windows and establishment of a feedback. These windows were bonded to the tubes by picein or glyptal, or were simply held by optical contact.

We investigated longitudinal-discharge systems, i.e., systems in which the voltage drop occurred along the tube axis. In most of the experiments the electrodes were cold and they were sealed into side branches at the tube ends. Thoriated tungsten electrodes, for example those taken from IFP-2000 lamps, were usually employed in the quartz tubes. No special investigation was made of the influence of the electrode shape and material on the stimulated emission parameters. However, the stimulated radiation power emitted from some systems (this was true, for example, of infrared stimulated emission from nitrogen) depended strongly on the cathode material.

The discharge conditions could be varied easily because we normally used tubes connected to a vacuum system via suitable valves. We employed a glass vacuum system with a TsVL-100 oil diffusion pump. This system made it possible to evacuate the discharge tubes to pressures of the order of 10^{-5} Torr. Before the measurements the discharge tubes were outgassed by striking a discharge in an inert gas (usually in neon). Such a conditioning followed by repeated evacuation ensured that the results of subsequent experiments were stable. The degree of outgassing could be judged roughly by the color of the discharge in neon. A more sensitive measure was provided by a visual examination of the neon discharge spectrum obtained using a UM-2 monochromator. Molecular bands could be observed in the discharge spectrum if the tube was poorly outgassed. A further check was sometimes made by observing some known laser emission or superradiance lines very sensitive to impurities.

We investigated a large number of different gases. We used only spectroscopically pure inert gases provided from "high-purity" reservoir. Hydrogen was admitted through a palladium leakage valve, which ensured that the gas was highly pure. In the pressure range 10^{-2}–10^{-1} Torr we used a thermocouple manometer calibrated for the investigated gas. Gases were mixed, when necessary, either in the vacuum system or directly in the discharge tube using the known volumes of the different parts of the vacuum system.

§ 6. Systems for Cooling Working Gases

Our investigations demonstrated (Chaps. I and III–V) that the gain due to molecular transitions increased considerably when the working gas was cooled. Consequently, many experiments with gases such as nitrogen and carbon monoxide were carried out at low temperatures right down to liquid nitrogen temperature (78°K). Various cooling systems were employed. One of them is shown schematically in Fig. 12a. Practically the whole of the discharge tube was surrounded by two fused quartz jackets. The outer vacuum jacket was evacuated by the diffusion pump and was used to provide thermal insulation; the inner jacket was used for cooling purposes and either it was filled with liquid nitrogen or cold nitrogen vapor was blown through it (this vapor was produced by evaporation of liquid nitrogen in an ASD-15 Dewar flask provided with a special heater). In the latter case, we could regulate the rate of evaporation of nitrogen and thus vary continuously the temperature of the discharge tube between room temperature and $\sim 100°K$.

In a different cooling system a discharge tube was in a special plastic-foam bath containing liquid nitrogen. The system was simple to construct but it was inconvenient to use with narrow tubes because the deformation of the plastic-foam bath due to its cooling was consider-

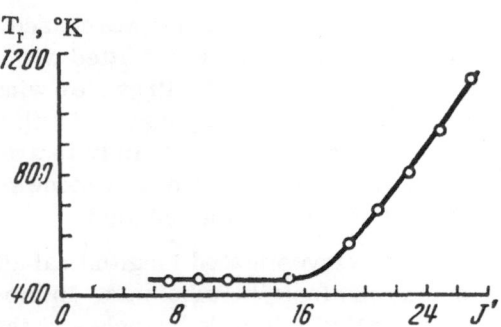

Fig. 13. "Rotational temperature" plotted as
a function of the rotational quantum number J,
deduced from the continuous $0-0$ band in the
2^+ system of nitrogen.

able and it could bend the discharge tube and thus cause misalignment of the resonator relative
to the tube axis. A system of this kind could be provided with a cover, and once again blowing
of cold nitrogen could be used to vary smoothly the tube temperature.

In all the experiments with cooled tubes we measured the temperature of the outer wall
of the tube. This temperature was determined with a copper−constantan thermocouple bonded
to the outer wall of the tube at the center of the working zone. Moreover, various methods
were employed in estimating the gas temperature in the tube. In particular, an attempt was
made to measure the rotational temperature deduced from the spontaneous radiation bands in
the 2^+ nitrogen system [59]. Numerous measurements demonstrated that the distribution of
the population in the rotational structure could not be described by a single temperature over
a wide range of J. Results of one such measurement are plotted in Fig. 13, which demonstrates
that high values of J apparently corresponded to higher rotational temperatures. However, at
relatively low values of J (within the region where the stimulated emission was observed) the
distribution of the population between the rotational sublevels could be described by a definite
temperature. This temperature agreed, within the limits of the experimental error, with the
gas temperatures estimated by other methods (for example, deduced from the breakdown
voltage of the discharge gap).

All the estimates obtained indicated that, under typical experimental conditions (when
the pulse repetition frequency was 2 Hz), the gas temperature in the tube was approximately
1.3 times higher than the temperature of its outer wall.

§ 7. Mirrors and Resonators. Superradiance and Laser
Emission

We used mirrors with metal (silver, aluminum) and multilayer dielectric coatings; their
reflection maxima were located in different parts of the spectrum. The mirrors were usually
spherical (radius of curvature 2 m) and were located in nearly confocal positions. Departures
from the general rule and also some specific parameters of the mirrors will be mentioned at
suitable points in this paper.

A special feature of pulse gas lasers was that high values of the gain were readily
attained. For some of the investigated lasers the gain was so high that stimulated radiation
could easily be observed without mirrors or with just one mirror. When the single-pass
gain was sufficiently high, the intensity of the stimulated radiation increased so much that we
readily observed directional radiation at the investigated line and the divergence was set by
the geometric dimensions of the tube. In this case the radiation emitted from long and narrow
discharge tubes resembled that emitted from a conventional laser with mirrors although the
divergence was somewhat higher. Sometimes this was observed only in the presence of one
mirror behind the tube, i.e., when the gain was due to two passes through the tube. This
operation regime was called superradiant and the radiation itself superradiance (the corre-

sponding German term is superstrahlung); it should be noted that the term superluminescence was sometimes used in the literature. Superradiance was characterized not only by directionality but also by a sharp increase in the intensity of the emitted line which was usually several orders of magnitude higher (along the tube axis) than the intensity of the corresponding spontaneous emission line. The duration of superradiance pulses was usually (this was true of our experiments) much shorter than the duration of the spontaneous radiation pulses. A criterion of the appearance of superradiance was the dependence of the line intensity on the experimental conditions (discharge current, gas pressure, etc.), which was very different from the corresponding dependence exhibited by a spontaneous line; under certain conditions, the intensity of the superradiance line rose extremely rapidly. Moreover, only strong rotational branches were observed in superradiance. All these features made it possible to identify superradiance quite reliably (in most cases).

We used the term laser emission for the case when stimulated radiation was observed only in the presence of an optical resonator (mirrors) with a considerable feedback. From the point of view of population inversion, the laser emission and superradiance regimes differed only in respect of the value of the gain. However, there was an important difference in respect of the nature of the radiation: in the presence of a resonator the radiation had a mode structure governed by the type and dimensions of the resonator [67]. The nature of superradiance was independent of the external mirrors and was governed by the properties of the active medium itself.

CHAPTER III

LASER EMISSION AND SUPERRADIANCE IN SECOND POSITIVE (ULTRAVIOLET) SYSTEM OF NITROGEN BANDS

§ 1. Characteristics of Laser Emission

Investigations of ultraviolet laser emission from molecular nitrogen were carried out mainly in relatively narrow tubes because the output power fell rapidly when the tube diameter was increased. The results reported below were obtained using a tube whose internal diameter was 3 mm and the active length was 90 cm. This tube had quartz optical-grade windows oriented at the Brewster angle and cold Duralumin electrodes separated by a gap of 100 cm. The voltage across the working capacitor was up to 55 kV.

The apparatus described in the preceding chapter was used to determine the average laser emission power as a function of the nitrogen pressure and the voltage across the capacitor. Figure 14 shows the dependences of the average (total for all the lines) laser emission power on the nitrogen pressure obtained for a constant voltage across the capacitor (44 kV) and on the voltage for a constant pressure (3 Torr). These dependences were obtained under normal conditions without cooling. It is clear from Fig. 14 that the laser emission was observed in a fairly wide range of pressures from a fraction of a torr to several torr. At a given voltage there was an optimal nitrogen pressure at which the laser emission power reached its maximum value. In this case this pressure was 2.5 Torr when the voltage was $V = 44$ kV. It is clear from Fig. 14b that at a fixed pressure the average power increased almost linearly with the voltage.

Measurements indicated that the optimal pressure increased with the voltage. The relationship between this optimal pressure and the voltage is shown in Fig. 15a. Clearly, the optimum always corresponded to a constant value of the ratio $\gamma (V - V_0)/N$. Similar results were obtained also for the limits of the range of conditions under which laser emission was

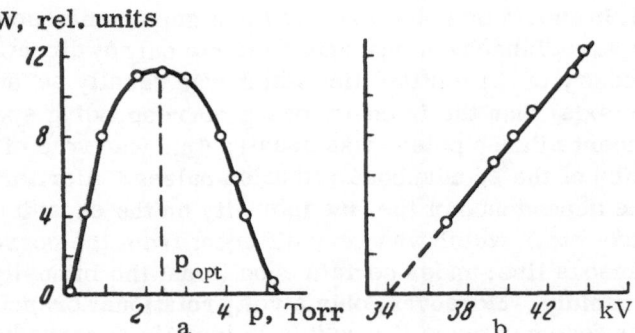

Fig. 14. Dependences of the power of the laser emission in the 2^+ nitrogen system on the gas pressure for a constant voltage V = 44 kV (a) and on the voltage across the working capacitor at a constant pressure p = 3 Torr at T = 320°K (b).

observed. It is clear from Fig. 15a that the higher the nitrogen pressure, the higher the voltage at which laser emission was observed. The dependence of the average total power on the voltage for a constant value of $\gamma = \gamma_{opt}$ is plotted in Fig. 15b.

Under our experimental conditions the ultraviolet laser emission from molecular nitrogen appeared at voltages of about 25 kV [14-16]. At the time this result was obtained it was of considerable practical importance because up to then the ultraviolet laser emission from nitrogen was observed only at voltages of 100-150 kV, which were inconvenient in practice.

Measurements carried out using an FÉU-39 photomultiplier demonstrated that the laser emission appeared at the very beginning of the high-current phase of the excitation pulse. The duration of the laser emission pulses was about 20 nsec or less (the measurements were carried out using the apparatus with a time resolution of this order of magnitude). The peak laser emission power reached several kilowatts and was considerably (by more than two orders of magnitude) higher than that reported in [3, 4]. It is important to note that the highest output power was obtained using mirrors whose transmission was about 40% in the working part of the spectrum; this was evidence of a high gain achieved in this laser system. Moreover, it was also possible to observe stimulated radiation in the superradiant regime. When one mirror was used, the output radiation was similar to that emitted by a conventional laser with resonator and the power was fairly high (0.3 kW). To the best of our knowledge, this was the first observation of superradiance due to molecular transitions. Roughly at the same time, Leonard [20] observed superradiance in the 2^+ nitrogen system as a result of much more powerful excitation in a transverse discharge.

Fig. 15. Power characteristics of the laser emission in the 2^+ system of N_2 at T = 320°K. a) Range of existence of laser emission and optimal conditions: 1) upper limit, $\gamma = \gamma_{max}$; 2) optimum, $\gamma = \gamma_{opt}$; 3) lower limit, $\gamma = \gamma_{min}$. b) Dependence of the average total laser emission power on the voltage for $\gamma = \gamma_{opt}$.

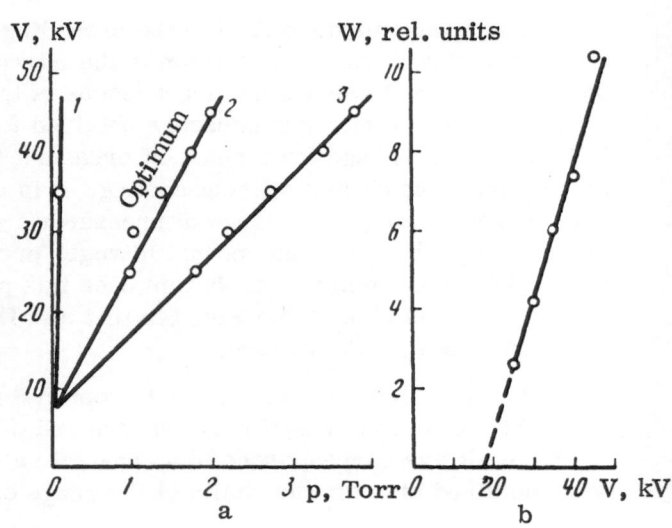

§2. Influence of the Gas Temperature

on Laser Emission

An investigation of the influence of temperature on the laser emission was started when certain special features were observed in the behavior of a laser system utilizing the 2^+ band system of nitrogen. It was found that during the first moments after the beginning of a discharge the output power did not reach immediately its steady-state value but took a fairly long time to rise. Figure 16 shows typical time dependences of the average output power measured from the beginning of the discharge. The initial region of rapid rise was, as established experimentally, due to the slow response of the detector. It is clear from Fig. 16 that it took several minutes to establish the steady-state power. At pressures close to the optimal value the power rose monotonically, whereas below the optimal pressure the steady-state value was approached after passing through a maximum.

Measurements of the temperature of the outer wall of the discharge tube indicated that the final temperature (chain curve in Fig. 16) was established in approximately the same time as the steady power. Therefore, it was natural to attribute this effect to the heating of the gas. Our systems had a large ballast volume connected to the discharge tube. This volume was much greater than the volume of the tube where the gas was heated. Therefore, the pressure in the system remained practically constant and only the gas density N varied and this density was related to temperature by the well-known expression $N = p/kT$. Thus, during the establishment of a discharge the temperature and density of the gas both varied with time.

The influence of temperature on the laser emission was investigated in a direct experiment in which the discharge tube was forced-cooled with liquid nitrogen. Figure 17a shows the dependence of the average total laser emission power on the nitrogen pressure obtained at two temperatures: 330 and 80°K. Figure 17b shows the same curves but as a function of the

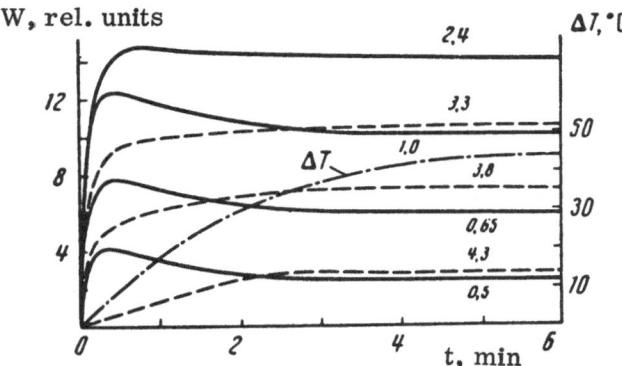

Fig. 16. Rise of the laser emission power with time for V = 44 kV and pulse repetition frequency 3 Hz; the numbers alongside the curves give the pressure in torr. The continuous curves correspond to pressures below the optimal value (2.5 Torr) and the dashed curves to those above the optimal pressure; the chain curve demonstrates the rise of the temperature of the outer wall of the tube due to the heating by the discharge.

Fig. 17. Power of the laser emission in the 2^+ nitrogen system at 80°K (curves denoted by 1) and 330°K (curves denoted by 2), plotted as a function of the pressure p (a) and of the working gas density N_0 (b).

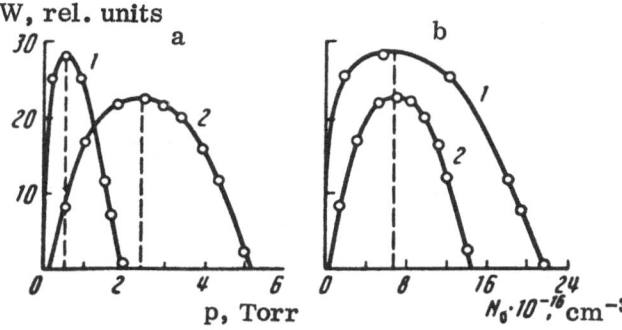

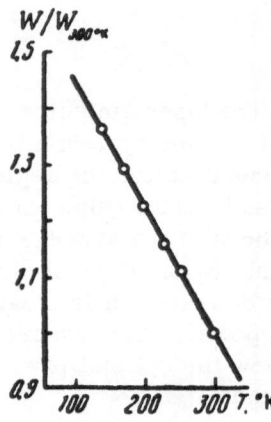

Fig. 18. Relative change in the total ultraviolet laser emission power with the gas temperature.

gas density. It is clear that at the two temperatures employed the maximum laser emission power was reached at different pressures but at the same density of nitrogen molecules, which was $N_{opt} = 7 \cdot 10^{16}$ cm^{-3}.

This result enabled us to explain easily the observed time dependence of the laser emission power. Since heating of the gas in the discharge tube reduced the density, the laser emission power in the range $N > N_{opt}$ should increase smoothly to its steady-state value, whereas in the range $N_{opt} < N$ the approach to the steady-state value should be accompanied by a fall of the density. However, a more detailed investigation of this process indicated that the influence of temperature did not reduce to the self-evident change in the density. This was manifested particularly clearly in measurements carried out at two very different temperatures. In particular, we found (Fig. 17) that at a fixed gas density the laser emission at 80°K was characterized by a higher power than at 330°K.

The temperature dependences of the output power were investigated (in a wide range of temperatures) in a special experiment in which the tube temperature was lowered smoothly from room temperature to that of liquid nitrogen. We employed a laser tube whose internal diameter was 8 mm and whose active length was 90 cm; this tube was surrounded by a jacket through which cold nitrogen was blown (see Chap. II). Figure 18 shows the dependence of the average total laser emission power on the temperature of the discharge tube walls, obtained for a constant (optimal) nitrogen density. The resonator mirrors were the same throughout this series of measurements. Clearly, the laser emission power increased smoothly when the temperature was lowered. Cooling from 300 to 100°K increased the output power by a factor of about 1.5. Experiments indicated that cooling also increased strongly the gain. In particular, at liquid nitrogen temperature we could observe clear superradiance (without any mirrors) using a tube whose internal diameter was 3 mm, whereas at room temperature such superradiance was obtained only when one mirror was used. Moreover, we found that cooling altered the laser emission spectrum and, in particular, the gain and output power of weak stimulated emission bands increased strongly. These changes in the spectra will be discussed in greater detail in the next section.

§3. Laser Emission Spectra

Before our investigation there had been hardly any studies of ultraviolet laser emission from nitrogen. We carried out a detailed investigation of the laser emission spectra because the knowledge of these spectra was needed for the understanding of the laser emission mechanism and for practical applications. The spectra were determined under various conditions. The majority of them were obtained using discharge tubes whose internal diameters were 3 and 8 mm. The spectra were recorded under laser emission and superradiance conditions. The majority of the measurements were carried out using a pair of mirrors with multilayer dielectric coatings, which transmitted 42, 39, and 70% of the incident radiation at wavelengths of 3160, 3370, and 3580 Å, respectively. Some of these experiments were carried out using dielectric mirrors whose transmission was 1.5% in the 3100-3600 Å range. In some cases we used aluminized mirrors.

The rotational spectrum of the 2^+ bands of nitrogen consisted of a large number of very closely spaced lines (about 50 lines per 1 Å near the band edges). Consequently, certain difficulties were encountered in the measurements and we had to use high-resolution and large-dispersion instruments. We recorded the stimulated and spontaneous radiation spectra of the 2^+ band system with a DFS-13 spectrograph using diffraction gratings with 600 and 1200 lines/mm in different orders. The spectrum of the 0−0 band was photographed mainly with a linear dispersion of 0.94 Å/mm, whereas the 0−1 and 1−0 bands were recorded with dispersions of 2 and 0.5 Å/mm, respectively. The arc spectra of iron and titanium were used as the reference standards. The relative error in the wavelength determination was ±0.004 Å. The absolute error was mainly due to the standard spectrum and it amounted to ±(0.02-0.04) Å. In all measurements we used "specially pure" nitrogen supplied from metal cylinders.

Spontaneous emission spectra of the nitrogen bands were recorded together with the laser emission and superradiance spectra in order to help and refine the identification of lines. The correctness of this identification was checked by plotting the Fortrat curves.

Our experiments revealed that laser emission occurred in three bands of the 2^+ system of nitrogen:

Band $(v' \to v'')$	1—0	0—0	0—1
Band edge (Å, air). . .	3159	3371	3577

No other laser emission bands or lines were observed. In all cases the strongest band was 0−0. Superradiance in this band was observed even at room temperature if a narrow tube was used. The 0−1 band was considerably weaker and superradiance was observed only when the gas was cooled. The 1−0 band was even weaker: laser emission was observed only when the tube was cooled. By way of example, Fig. 19 shows schematically the laser emission spectrum in the region of the 0−0 band and a diagram of relative intensities of the observed lines.

The stimulated emission wavelengths obtained in our experiments are listed in Table 3. This table gives, for all three bands, the values of λ_{calc} and ν_{calc}, calculated from the tabulated energies of molecular levels [68]. The attributions of the observed lines to specific rotational transitions are given in the last columns of the table. They show the component of a multiplet (Ω) or rotational branch, and the component of the Λ doubling for each line as well as the rotational quantum number J" of the lower state. We observed in total 71 stimulated emission lines corresponding to 146 rotational transitions. The majority of these lines corresponded to the strongest band 0−0 (43 lines). Three of these lines could not be attributed in a definite manner. Since the publication of our results, measurements of the stimulated radiation spectra in the 0−0 band region were reported by other workers [17-19]. A comparison with these results is made in the next section.

Table 3 summarizes the situation in the sense that not all the lines were measured under the same experimental conditions. The majority of the lines in the spectra of the 0−0 and 1−0 bands were observed at a pressure of 0.5 Torr when the tube was cooled with liquid nitrogen, whereas the lines in the 0−1 band were obtained at a pressure of 1 Torr without cooling. The calculated wavelengths λ_{calc}, enclosed in brackets, were taken from the experimental paper [68] because of the lack of data which were needed in calculations.

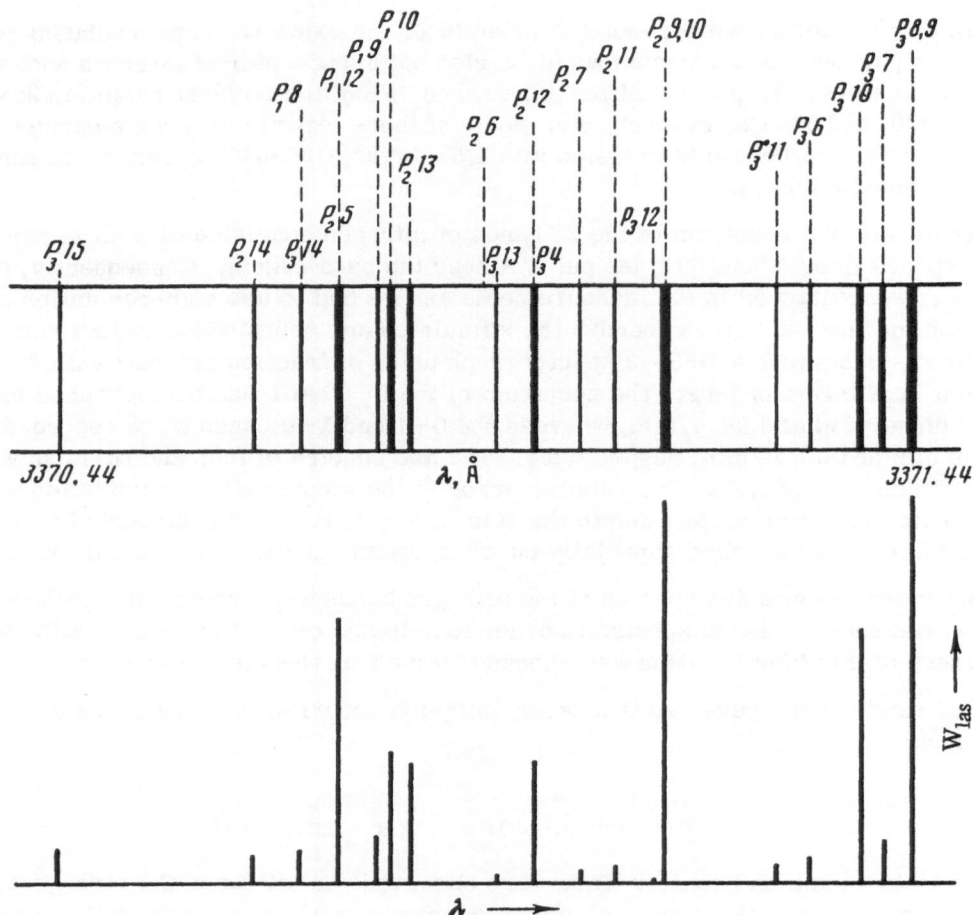

Fig. 19. Schematic representation of the 0−0 laser emission band in the 2$^+$ nitrogen system and diagrammatic representation of the line intensities at p = 1.4 Torr, V = 42.6 kV, T = 320°K.

Our experiments demonstrated that considerable changes occurred in the laser emission spectra when the tube was cooled. A redistribution of the intensities occurred between the rotational lines, and the maximum of the distribution shifted in the direction of lower values of J when the total power was fairly high. In the absence of cooling the highest stimulated emission power corresponded to lines with $J'_{max} \approx 8-9$, whereas when liquid nitrogen was used, the highest power corresponded to $J'_{max} \approx 5$. Cooling gave rise to new lines in the spectrum, and power increased for the band as a whole, particularly for the lines near the rotational distribution maximum. The cooling-induced rise of the laser emission power differed from band to band. For example, it was considerably greater for the 0−1 band than for the strongest 0−0 band. Clearly, the increase in the power was due to an increase in the gain. For example, in the case of the 0−1 band, which was very weak at room temperature, we observed even superradiance at liquid nitrogen temperature. Finally, cooling led to the discovery of a new 1−0 band which had not been observed before in the room-temperature laser emission spectra. Until very recently this was the shortest-wavelength band (3159 Å) in the molecular stimulated emission. A comparison of the laser emission and superradiance spectra demonstrated that when the power of superradiance was sufficiently high, its spectrum was identical with the laser emission spectrum.

TABLE 3. Rotational Laser Emission Structure in Various Bands of Second Positive System of Nitrogen (absolute error ±0.04 Å, relative error ±0.0036 Å)

Line No.	λ_{meas}, Å (in air)	λ_{calc}, Å (in air)	ν_{calc}, cm⁻¹ (in vacuum)	Rotational transition $\ell=0$	$\ell=1$	$\ell=2$	Transition No.
			0 — 0 band				
1	3371.438	3371.440	29652.43			$P_3 9^c$	1
		3371.439	29652.44			$P_3 8^c$	2
2	3371.401	3371.400	29652.78			$P_3 7^c$	3
3	3371.376	3371.380	29652.96			$P_3 10^c$	4
4	3371.320	3371.319	29653.49			$P_3 6^c$	5
5	3371.279	3371.281	29653.83			$P_3 11^c$	6
6	3371.185	3371.183	29654.67			$P_3 5^c$	7
7	3371.149	3371.154	29654.95		$P_2' 9^c$		8
		3371.150	29654.98		$P_2 10^c$		9
8	3371.136	3371.144	29655.03		$P_2' 10$		10
		3371.142	29655.05			$P_3' 12^c$	11
9	3371.120	3371.120	29655.24		$P_2 8^c$		12
		3371.119	29655.25		$P_2 8$		13
10	3371.089	3371.094	29655.47		$P_2 11$		14
		3371.090	29655.51		$P_2 11^c$		15
11	3371.041	3371.042	29655.93		$P_2' 7^c$		16
12	3370.996	3370.997	29656.33				17
		3370.996	29656.34			$P_3' 4^c$	18
13	3370.943	3370.950	29656.74		$P_2 12^c$		18
						$P_3 13^c$	19
14	3370.931	3370.939	29656.84		$P_2 9^c$		20
		3370.934	29656.88		$P_2 6$		21
15	3370.848	3370.855	29657.58		$P_2 13$		22
		3370.851	29657.61		$P_2 13^c$		23
		3370.847	29657.65	$P_1 10$			24
		3370.839	29657.75	$P_1 11^c$			25
16	3370.820	3370.824	29657.85	$P_1' 10^c$			26
		3370.812	29657.95	$P_1' 11$			27
17	3370.806	3370.809	29657.98	$P_1 9^c$			28
18	3370.762	3370.766	29658.36		$P_2' 5^c$		29
		3370.764	29658.38	$P_1' 12^c$			30
19	3370.720	3370.724	29658.73	$P_1' 8^c$			31
20	3370.717	3370.719	29658.77				32
21	3370.666	3370.669	29659.21			$P_3' 14^c$	33
22	3370.628	3370.627	29659.58	$P_1 7^c$			34
23	3370.567	3370.579	29660.00		$P_2 14^c$		35
24	3370.560	3370.559	29660.18	$P_1 14$	$P_2 4^c$		36
25	3370.530	3370.539	29660.36	$P_1' 14^c$			37
26	3370.479	3370.485	29660.83	$P_1 6$			38
27	3370.473	3370.473	29660.94	$P_1' 6^c$			39
28	3370.444	3370.433	29661.11		$P_2' 15^c$		40
		3370.444	29661.19			$P_3 15^c$	41
29	3370.372	3370.366	29661.88	$P_1' 15$			42
30	3370.326	3370.331	29662.19		$P_2' 3^c$		43
31	3370.305	3370.306	29662.41	$P_1 5^c$			44
32	3370.170	[3370.174]	—	$P_1 16$			45
33	3370.137	[3370.137]	—	$P_1' 16^c$			46
34	3370.085	3370.089	29664.32	$P_1 4$			47
		3370.080	29664.40	$P_1' 4^c$			48
35	3369.848	3369.846	29666.46	$P_1 3^c$			49
		3369.844	—		$P_2 17$		50
36	3369.769	—	—	—	—	—	51
37	3369.555	[3369.551]	—	—		$Q_3 2^c$	52
38	3369.259	3369.256	29671.65	$P_1' 1$	$Q_2 1^c, Q_2' 1$		53
		3369.253	29671.68	$P_1 1^c$			54
39	3368.435	[3368.428]	—	$P_1 21$		$P_3' 20^c$	55
40	3366.915	3366.915	29692.28			$R_3' 4^c$	56
41	3365.481	3365.482	29704.92			$R_3' 6^c$	57
42	3365.425	—	—		—	—	58
43	3364.906	3364.910	29709.97		$R_2' 7^c$		59
		3364.909	29709.98	$R_1' 7$			60

TABLE 3. Continued

Line No.	λ_{meas}, Å (in air)	λ_{calc}, Å (in air)	ν_{calc}, cm⁻¹ (in vacuum)	Rotational transition Ω=0	Ω=1	Ω=2	Transition No.
				0−1 band			
1	3576.949	3576.950	27948.80			$P_3'8^{\circ}$	1
		3576.948	27948.82			$P_3 7^{\circ}$	2
2	3576.891	3576.891	27946.26			$P_3'6^{\circ}$	3
3	3576.616	3576.620	27951.38		$P_2 8^{\circ}$		4
		3576.615	27951.42		$P_2 9,\ P_2'8$		5
4	3576.612	3576.612	27951.44			$P_3 11^{\circ}$	6
		3576.611	27951.45		$P_2'9^{\circ}$		7
5	3576.468	3576.471	27952.54		$P_2 6^{\circ}$		8
		3576.468	27952.57		$P_2''6$		9
6	3576.235	3576.249	27954.28	$P_1 9^{\circ}$			10
		3576.241	27954.34	$P_1 10$			11
7	3576.173	3576.192	27954.72	$P_1'8^{\circ}$			12
		3576.185	27954.78	$P_1 11^{\circ}$			13
8	3576.111	3576.119	27955.29	$P_1 7^{\circ}$			14
		3576.109	27955.37		$P_2'4$		15
9	3576.095	3576.107	27955.39	$P_1'7$	$P_2'4^{\circ}$		16
10	3575.789	3575.810	27957.71	$P_1'5^{\circ}$			17
		3575.798	27957.80	$P_1'5$			18
11	3575.459	—	—	—	—		19
				1−0 band			
		3159.218	31644.64			$P_3 9^{\circ}$	1
		3159.203	31644.79			$P_3 10^{\circ}$	2
1	3159.19	3159.175	31645.07			$P_3'8^{\circ}$	3
		3159.166	31645.16			$P_3 11^{\circ}$	4
2	3159.11	3159.118	31645.64			$P_3 7^{\circ}$	5
		3159.092	31645.90			$P_3 12^{\circ}$	6
3	3159.00	3159.014	31646.68			$P_3'6^{\circ}$	7
		3158.983	31646.99			$P_3 13^{\circ}$	8
		3158.938	31647.44		$P_2 10^{\circ}$		9
		3158.933	31647.49		$P_2'10$		10
		3158.931	31647.51		$P_2 11$		11
4	3158.91	3158.928	31647.54		$P_2'11^{\circ}$		12
		3158.900	31647.82		$P_2 9$		13
		3158.892	31647.91		$P_2 12^{\circ}$		14
		3158.880	31648.02		$P_2'9^{\circ}$		15
		3158.874	31648.09			$P_3 5^{\circ}$	16
		3158.853	31648.30			$P_3'14^{\circ}$	17
		3158.842	31648.41		$P_2 8^{\circ}$		18
5	3158.83	3158.840	31648.43		$P_2'8$		19
		3158.820	31648.63		$P_2'13^{\circ}$		20
		3158.818	31648.65		$P_2 13$		21
6	3158.74	3158.765	31649.18		$P_2 7$		22
		3158.744	31649.39	$P_1'7^{\circ}$			23
7	3158.70	3158.716	31649.67		$P_2 14^{\circ}$		24
		3158.689	31649.94			$P_3'4^{\circ}$	25
		3158.648	31650.35			$P_3 15^{\circ}$	26
		3158.636	31650.47	$P_1 12$			27
		3158.634	31650.49	$P_1 11^{\circ}$			28
		3158.625	31650.58		$P_2 6^{\circ}$		29
		3158.623	31650.60		$P_2''6$		30
8	3158.61	3158.614	31650.69	$P_1'12^{\circ}$			31
		3158.612	31650.71	$P_1'11$			32
		3158.604	31650.79	$P_1 13^{\circ}$			33
		3158.603	31650.80	$P_1 10$			34
		3158.589	31650.94	$P_1'13$			35
		3158.582	31651.01	$P_1'10^{\circ}$			36
		3158.571	31651.12		$P_2'15^{\circ}$		37
		3158.539	31651.44	$P_1 14$			38
		3158.536	31651.47	$P_1 9^{\circ}$			39
9	3158.53	3158.519	31651.64	$P_1'9$			40
		3158.517	31651.66	$P_1'14^{\circ}$			41
		3158.461	31652.22		$P_2''5$		42
		3158.458	31652.25		$P_2'5^{\circ}$		43
10	3158.44	3158.440	31652.43	$P_1 15^{\circ}$			44
		3158.438	31652.45	$P_1 8$			45
		3158.428	31652.55	$P_1'15$			46
		3158.425	31652.58	$P_1'8^{\circ}$			47

TABLE 3. Continued

Line No.	λ_{meas}, Å (in air)	λ_{calc}, Å (in air)	ν_{calc}, cm⁻¹ (in vacuum)	Rotational transition $\Omega=0$	$\Omega=1$	$\Omega=2$	Transition No.
			1—0 band				
11	3158.32	3158.315	31653.69	$P_1 7^c$			48
		3158.305	31653.79	$P_1' 7$			49
12	3158.27	3158.287	31653.97		$P_2' 4$		50
		3158.272	31654.12		$P_2 4^c$		51
13	3158.16	3158.166	31655.18	$P_1 6$			52
		3158.152	31655.32	$P_1' 6^c$			53
14	3158.03	3158.039	31656.45		$P_2' 3^c$		54
		3158.034	41656.50		$P_2 3$		55
15	3157.98	3157.987	31656.97	$P_1 5^c$			56
		3157.976	31657.08	$P_1' 5$			57
		3157.783	31659.02		$P_2 2$		58
16	3157.78	3157.782	31659.03	$P_1 4$			59
		3157.773	31659.12	$P_1' 4^c$			60
		3157.772	31659.13		$P_2 2^c$		61
17	3157.56	3157.544	31661.42	$P_1' 3$			62
		3157.520	31661.66	$P_1 3^c$			63

Note. Letter "c" is used to denote the chief (strong components) of Λ doubling (lines due to the ortho modification of nitrogen); a prime is used for those transitions whose upper level has a higher energy in the P_2 branch or a lower energy in the P_1 and P_3 branches (in the Λ splitting).

The distribution of the power between the separate rotational lines in a band was fairly complex. We shall consider only typical examples. Figure 20 shows the distribution of the power between the P_1, P_2, and P_3 rotational branches of the 0—0 band. These spectra were obtained without cooling by applying a voltage of 42.6 kV at T = 320°K to tubes filled with a gas at pressures of 1.4, 2.0, and 2.6 Torr. The pressure of 2.0 Torr was nearly optimal. It is clear from Fig. 20 that the measured distributions could not be described easily by smooth curves. Nevertheless, each rotational branch had two distributions of the primed (in Table 3) and unprimed components, represented by the continuous and dashed curves, respectively.* In each distribution we included transitions only due to the ortho modification: For the primed components in the P_1 and P_3 branches these were the transitions with even values of J, whereas in the P_2 branch they were the transitions with odd values of J; the converse was true of the unprimed components. Moreover, we found that in each distribution, lines either decreased in intensity or disappeared completely at intervals of $\Delta J = 4$: in the case of the primed components this was true of the lines with J″ = 6, 10, and 14 in the P_1 and P_3 branches and with J″ = 7, 11, and 15 in the P_2 branch; for the unprimed components this was true for lines with J″ = 7 or 11 in the P_1 and P_3 branches and with J″ = 8 or 12 in the P_2 branch.

Figure 21 shows the corresponding distributions for superradiance obtained by cooling the discharge tube with liquid nitrogen and applying approximately the same voltage (45.2 kV) and a pressure of 0.5 Torr, which corresponded to the same gas density as at a pressure of 2 Torr without cooling. The results were basically the same. The differences were observed

---

* In the case of the even (g) electronic states a prime was used for the positive rotational sublevels with even values of the rotational number K and negative sublevels with odd values of K; in the case of the odd (u) electronic states a prime was used for positive sublevels with odd values of K and negative sublevels with even values of K.

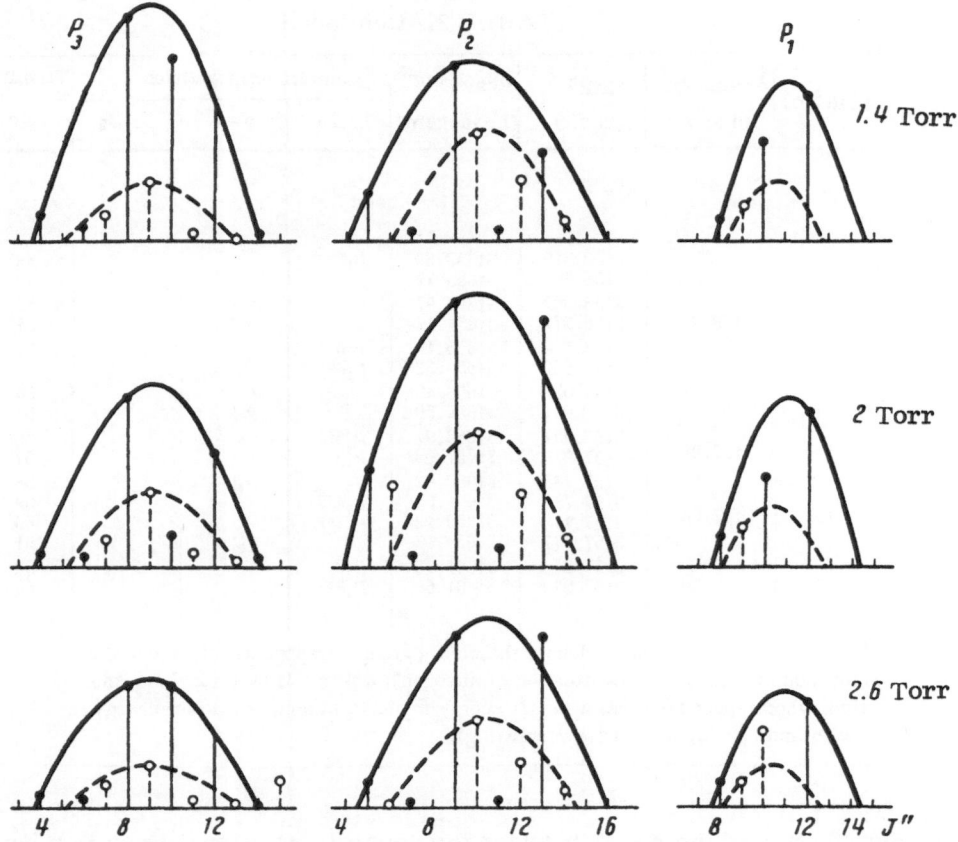

Fig. 20. Distribution of the laser emission power between the rotational components of the three main branches P_1, P_2, and P_3 of the 0−0 band in the 2^+ nitrogen system shown for different pressures (p = 2 Torr is the optimal pressure). The continuous curves represent the lines starting from the component of Λ doublets with lower (branches P_1 and P_3) and higher (branch P_2) energies. The dashed curves correspond to lines with the opposite situation.

in the relationships between the branches, in reduction of J_{max}, and in anomalously high intensities of some of the lines.

§ 4. Discussion of Results

We shall first consider the laser emission spectra. The emission spectra of the 2^+ system of nitrogen were investigated by us and also in [17-19], but only for the 0−0 band. Therefore, we compared our results with those published elsewhere only for the 0−0 band. This comparison is made in Table 4. It gives the measured values of the wavelengths and the attribution of the lines to specific rotational transitions given in the various papers. In [17] the wavelengths were measured in vacuum. These wavelengths were converted to the wavelengths in air so as to facilitate a direct comparison with our results. Different workers observed and measured properties of different numbers of laser emission lines. The greatest number of lines was obtained in our investigation and the smallest in [17]. We not only observed the $P_1 13^c$ line reported in [17, 18], but also the line at 3371.418 Å, observed only in [19] and not attributed to any definite transition. The identification, made in [18], of the 3366.9098 Å line as $R_2 4$ was, in our opinion, incorrect. It was more likely to be due to the $R_3 4$ transition.

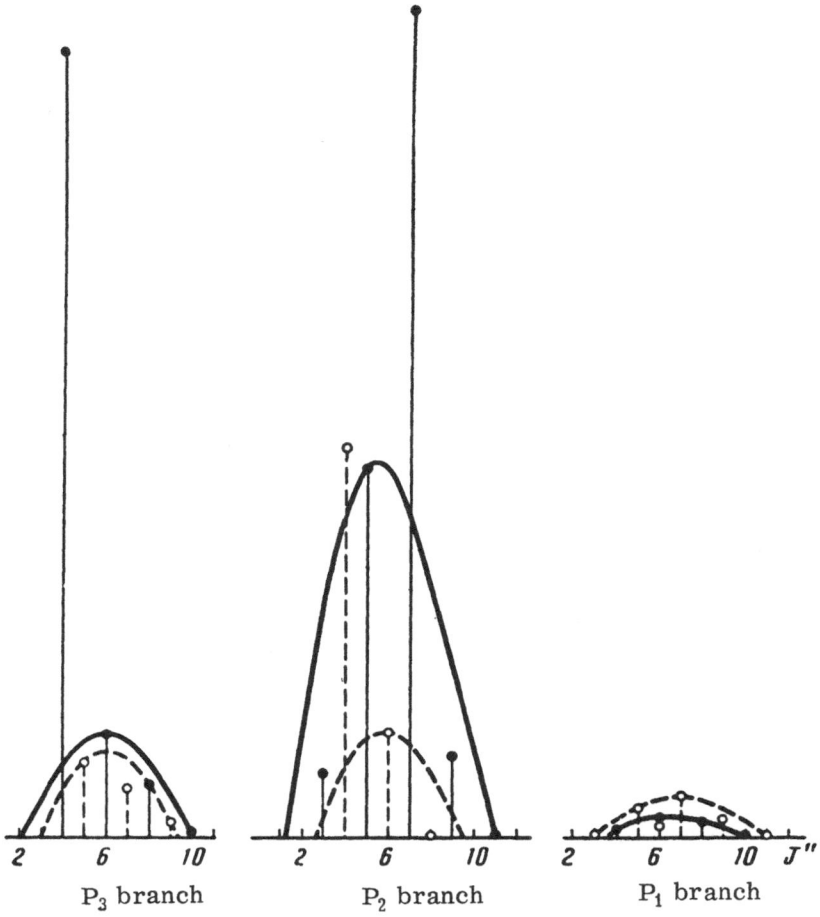

Fig. 21. Distribution of the superradiance power between the rotational components of the P_1, P_2, and P_3 branches of the $0-0$ band in the 2^+ nitrogen system (p = 0.5 Torr, V = 45.2 kV, T = 80°K). The designations have the same meaning as in Fig. 20.

Table 4 gives only those of the discovered lines which were also observed by other workers (a full list of the stimulated emission lines in the $0-0$ band can be found in Table 3). In those cases when information was available, we included the estimates of the error in the absolute and relative measurements of the wavelengths.

It is clear from Table 4 that, within the limits of the experimental error, the measured wavelengths agreed completely and there was also complete agreement in the attribution of the lines to specific rotational transitions and specific values of J. There were only a few discrepancies between the line attributions. One should mention particularly the excellent agreement between our results and those reported in [19]. This was due to the fairly high precision of the determination of the relative line positions, because in both cases the absolute values of the wavelengths were measured relative to the $P_3 8,9$ line whose wavelength was taken from [68] ($\lambda = 3371.437$ Å). However, it should be noted that the value adopted in our measurements was 0.001 Å higher. This shift was introduced to obtain a better agreement with the calculated wavelengths of all the other laser emission lines. The absolute wavelengths taken from [18] were approximately 0.01 Å less than in the other investigations.

TABLE 4. Wavelengths (Å, in air) of Laser Emission Lines in 0−0 Band of 2⁺ Nitrogen System (comparison of results obtained by different authors)

Our results	[17]	[18]	[19]
Absolute error ±0.04 Å Relative error ±0.003 Å		±0.001 Å ±0.0003 Å	
3371.438 $P_3 9^c$, $P_3' 8^c$	3371.44 $P_3 9, 9$	3371.4289 $P_3 8$ 3371.4215 $P_3 9$	3371.437 $P_3 9,8$
—			3371.418 —
3371.401 $P_3 7^c$	3371.41 $P_3 7$	3371.3920 $P_3 7$	3371.404 $P_3 7$
3371.376 $P_3' 10^c$	3371.39 $P_3 10$	3371.3658 $P_3 10$	3371.379 $P_3 10$
3371.320 $P_3' 6^c$	3371.31 $P_3 6$	3371.3070 $P_3 6$	3371.319 $P_3 6$
3371.279 $P_3 11^c$	3371.27 $P_3 11$	3371.2655 $P_3 11$	3371.278 $P_3 11$
3371.185 $P_3 5^c$	3371.20 $P_3 5$	3371.1722 $P_3 5$	3371.183 $P_3 5$
3371.149 $P_2 9^c$ $P_2' 10^c$	3371.16 $P_2 9$	3371.1385 $P_2 9$ $P_3 12$	3371.148 $P_2 9$ $P_3 12$
3371.136 $P_2' 10$ $P_3' 12^c$	3371.14 $P_2 8, 10$	3371.1232 $P_2 10$	3371.138 (?) $P_2 11$
3371.120 $P_2 8^c$, $P_2' 8$		3371.1140 $P_2 8$	
3371.089 $P_2 11$, $P_2' 11^c$	3371.08 $P_2 11$	3371.0770 $P_2 11$	3371.087 —
3371.041 $P_2' 7^c$	3371.04 $P_2 7$	3371.0369 $P_2 7$	3371.047 $P_2 7$
3370.996 $P_3 4^c$ $P_2' 12^c$	3370.99 $P_2 12$ $P_3 4$	3370.9848 $P_2 12$	3370.993 $P_2 12$ $P_3 4$
3370.941 $P_3 13^c$	—	—	—
3370.931 $P_2 6^c$, $P_2' 6$	3370.93 $P_2 6$	3370.9190 $P_2 6$	3370.928 $P_2 6$, $P_3 13$
3370.848 $P_2 13^c$, $P_2' 13^c$ $P_1 10$ $P_1' 11^c$	3370.84 $P_1' 10$ $P_1 9$ $P_1 11$	—	3370.845 $P_2 13$ $P_1 11$
3370.820 $P_1' 10^c$	—	3370.8147 $P_1 10$ $P_1 11$ $P_2 13$	3370.822 $P_1 10$ $P_1 11$
3370.806 $P_1' 11$ $P_1 9^c$	—	3370.7981 $P_1 9$	3370.807 $P_1 9$
3370.762 $P_2' 5^c$ $P_1' 12^c$	3370.76 $P_1' 12$	3370.7533 $P_1 12$ $P_2 5$	3370.763 (?) $P_2 5$ $P_3 3$
3370.720 $P_1' 8^c$	3370.72 $P_1' 8$	3370.7128 $P_1 8$	3370.720 $P_1 8$, $P_3 14$
3370.717 $P_3' 14^c$	—	—	—
—	3370.66 $P_1 7, P_1 13$	3370.6761 $P_1 13$	—
3370.666 $P_2' 14^c$	—	3370.6587 $P_2 14$	3370.668 $P_2 14$
3370.628 $P_1 7^c$	—	3370.6138 $P_1 7$	3370.626 $P_1 7$
3370.560 $P_1 14$	—	(?) 3370.5544 $P_1 14$	3370.532 $P_1 14$
3370.530 $P_1' 14^c$	—	—	—
3370.473 $P_1' 6^c$	—	3370.4660 $P_1 6$	—
3370.444 $P_2' 15^c$ $P_3 15^c$	3370.43 $P_1 15$	(?) 3370.4316 $P_2 15$	3370.438 $P_2 15$ $P_3 15$
3370.305 $P_1' 5^c$	—	3370.2944 $P_1 5$	—
3370.085 $P_1 4$, $P_1' 4^c$	—	3370.0755 $P_1 4$	3366.912 $R_2 4$ $R_3 4$
3366.915 $R_3' 4^c$		(?) 3366.9098 $R_2 4$	
3364.906 $R_2' 7^c$ $R_1 7$	—	(?) 3364.9075 $R_1 7$	—

Note. The symbol "?" denotes lines whose identification is, in our opinion, doubtful. See also footnote in Table 3.

Thus, a comparison of the results obtained in different investigations of the rotational structure of the 0−0 laser emission band indicated that the precision of the measurements was sufficiently high and the attribution of the laser emission lines to specific transitions usually caused no difficulties. Since the measurements of the wavelengths of the laser emission lines in the other bands (0−1 and 1−0) were carried out in our study using the same apparatus and under similar conditions to those adopted in the measurements of the 0−0 lines, we could conclude that the former were identified reliably.

The nature of the observed laser emission and superradiance spectra was basically in agreement with the calculations carried out on the assumption of direct excitation of the laser levels by electrons from the ground state of the molecule (Chap. I). In accordance with the calculations, the highest gain should be exhibited by the 0−0 band and the lowest by the 1−0 band. Moreover, in agreement with the calculations, the P branch predominated whereas the R branch was represented by just four lines in the strongest (0−0) band. This could be explained by the usual competition between the P and R transitions.

The distributions of the power in the rotational structure were not smooth. The origin of this type of distribution was not clear. Particularly, large anomalies were observed in the P_2 and P_3 branches. For example, in the case of superradiance in the P_3 branch we found that the P_34 line had an anomalously high power (Fig. 21). The upper rotational sublevel of the transition responsible for this line was in very exact resonance with the rotational sublevels of the other multiplet splitting components ($\Delta\nu < 3$ cm^{-1}). It is probable that rapid exchange of energy between closely spaced sublevels took place and the energy of these sublevels was transferred mainly to the P_34 line. We should then observe the relative reduction in the intensities of the P_26 and P_18 lines, which was confirmed experimentally. However, we were unable to explain in this way the other cases of an anomalous increase in the line power (for example, in the P_2 branch).

At first sight it seemed somewhat strange that the strongest branch was P_2 (see Figs. 20 and 21). It would seem that in the excitation of multiplet electronic states the energy should be distributed between the multiplet components in accordance with their statistical weights ($2\Omega + 1$), where Ω is the total angular momentum of the multiplet component in question. In this case the strongest branch should be P_3, for which the angular momentum is highest ($\Omega = 2$). The predominance of the P_2 branch found experimentally can be explained bearing in mind two points. On the one hand, the electron-impact excitation of the molecules occurs most effectively when there is no change in the projection of the spin angular momentum onto the internuclear axis. For the middle components ($\Omega = 1$) of the laser levels, i.e., for the states $C^3\Pi_u$ and $B^3\Pi_g$, responsible for the P_2 branch, the projection of its spin angular momentum onto the internuclear axis is zero. On the other hand, we may assume that the corresponding projection of the ground state $X^1\Sigma_g^+$, whose spin momentum is generally zero and which is the static point of the excitation of the active levels, is also zero. Thus, the excitation conditions are optimal for the P_2 branch. The other point to bear in mind is that an equilibrium between multiplet components is not established during stimulated emission pulses. In those cases when an equilibrium of this kind is established, the strongest branch should be P_3 and the weakest P_1.

It is shown in Chap. I that the nature of the distribution of the gain between the rotational components of a vibrational band is a criterion of the type of rotational relaxation (fast or slow). In other words, a comparison of the observed and calculated spectra makes it possible to establish whether a Boltzmann distribution of the populations between the vibrational sublevels with a temperature $T_r = T_{mol}$ is established in the time available or whether an intermediate case is realized. This method is most effective for very weak vibrational inversions. Experimental investigations of the 0−0 band (for which the calculations were made) indicated that it was characterized by high values of the vibrational inversion coefficient. Under these conditions the changes in the rotational structure were so slight that, within the limits of the experimental error, it was not possible to say anything definite about the nature of the rotational relaxation.

We shall now consider the characteristics of laser emission as a function of the gas density and the voltage applied to the capacitor. We can see from Fig. 15a that laser emission was observed in a certain range of voltages V and pressures p, limited by straight lines converging at the point $V = V_0$. Experiments indicated that the value of V_0 for our apparatus remained practically constant and equal to about 5-7 kV. As reported later, analogous dependences with similar values of V_0 were also obtained for laser emission from all the other lasers investigated by us. Our experiments demonstrated that the value of V_0 was a characteristic of our apparatus. It described the voltage drop on external (relative to the discharge tube) parts of the discharge circuit, including the circuit inductance, discharger, etc. This value included also the cathode drop in the discharge. Thus, we could regard $(V - V_0)$ as the initial voltage drop across the active part of the discharge. Since the points corresponding to the limits of the laser emission region and the optimal conditions fitted well linear dependences, we concluded that the optimal conditions and the limits corresponded to definite values of the ratio $(V - V_0)/p$. Investigations carried out at various temperatures indicated that the controlling factor was the dependence on the gas density and not on the gas pressure. Therefore, the optimal conditions and the limits of laser emission were characterized at all temperatures by the parameter $\gamma = (V - V_0)/N$, where N is the working gas density. This parameter is similar to the ratio E/p (see, for example, [69]), which governs the properties of the electron gas in a discharge and particularly the electron temperature or their average energy.

Thus, the results obtained indicated that the laser emission of lines in the 2^+ system of nitrogen was governed by the electron energy in the discharge. Hence, we concluded that laser emission was due to the interaction of the nitrogen molecules with electrons. The same conclusion followed from a comparison of the experimental and calculated parameters of the laser emission.

The laser emission of the 2^+ nitrogen bands was the first stimulated radiation due to electronic transitions in molecules to be studied as a function of the discharge temperature. A calculation of the distribution of the gain in the rotational band structure as a function of the gas temperature was also carried out for these bands (see Chap. I). The calculations were made on the assumption of the direct population of the active levels by electrons from the ground state of the molecules. It should be noted that the experimentally observed behavior of the laser emission from cooled tubes was in good agreement with the calculations. Thus, cooling shifted the power maximum in the direction of lower values of J and resulted in a strong rise of the gain manifested by the appearance of a clear superradiance in the 0−0 and 0−1 bands. Finally, cooling resulted in laser emission of a new band (1−0). The rise of the laser emission power varied from band to band. The rise was relatively small for the strongest (0−0) band (approximately a factor of 1.5 due to cooling with liquid nitrogen), whereas it was much higher for the 0−1 band. This was due to the fact that even at room temperature the 0−0 gain was high and close to saturation, whereas the 0−1 band approached saturation only as a result of cooling. It should be noted that we made no attempt to optimize the mirror transmission. A selection of the exit mirrors should increase the power still further.

We shall now discuss the population inversion mechanism responsible for stimulated transitions in the second positive system of nitrogen.

The N_2 molecule has been investigated more thoroughly than probably any other. Its properties have been studied in a large number of experimental and theoretical investigations. We shall now consider those characteristics of the nitrogen molecule which are relevant to the population inversion mechanism.

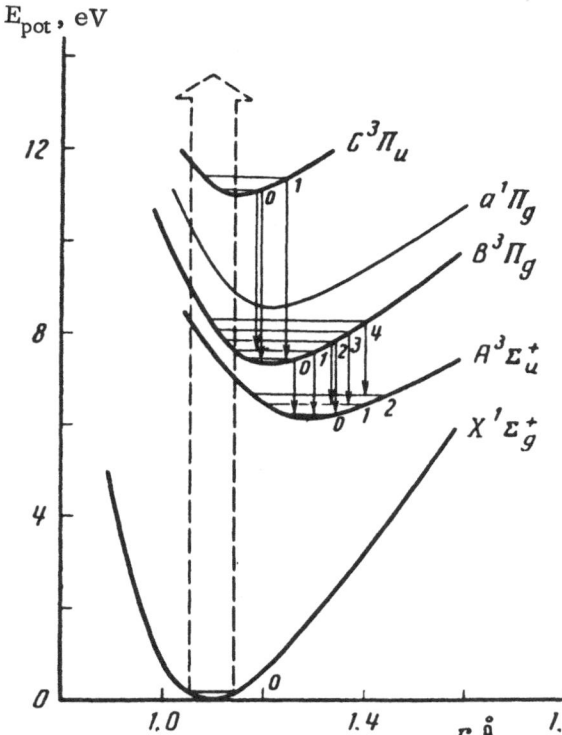

Fig. 22. Energy level scheme of the N_2 molecule.

Figure 22 shows schematically the potential curves of the N_2 molecule. It includes only those states which do or may participate in the stimulated emission of the 2^+ and 1^+ nitrogen systems. These potential curves were calculated in [70] by the Rydberg−Klein−Rice method. The spectral measurements demonstrated that the ultraviolet laser emission was due to the $C^3\Pi_u - B^3\Pi_g$ transition. The laser emission bands are identified by arrows in Fig. 22.

The $C^3\Pi_u$ state decays radiatively along two channels: 1) a strong transition along the active (laser) channel to the $B^3\Pi_g$ state; 2) a weak intercombination transition to the ground state $X^1\Sigma_g^+$ (Tanaka system). The latter transition is observed only under special experimental conditions [71]. The state C has only five vibrational levels. The levels with $v > 4$ predissociate. No cascades to the state C are observed experimentally.

The $B^3\Pi_g$ state decays radiatively only to the $A^3\Sigma_u^+$ state (first positive system). Infrared laser emission is observed as a result of this transition. It is discussed further in Chap. IV. The state B is the final state not only for transitions in the 2^+ system but also for transitions from the $D^3\Sigma_u^+$ state (fourth positive band system), from $C'^3\Pi_u$ (Goldstein−Kaplan bands terminate at vibrational levels with $v = 9$-13. Since we shall not be interested in the B-state levels with $v \geq 5$, we shall not discuss any further the Goldstein−Kaplan system.

Radiative decay of the metastable state $A^3\Sigma_u^+$ is very unlikely. Nevertheless, under special conditions transitions from this state to the ground state can be observed (Vegard−Kaplan system) [72]. Only one radiative system (1^+) terminates in the state A.

In discussing population inversion mechanisms we must consider particularly the electron-excitation cross sections of the laser states and the probabilities of radiative decay of these states. Since in the pulse systems of interest to us before the application of an excitation pulse practically all the molecules are in the lower vibrational level $v = 0$ of the ground state, we have to know the cross sections for the excitation from this level to various vibrational

TABLE 5. Excitation Cross Sections, Radiative Lifetimes,
and Rotational Constants of
$C^3\Pi_u$, $B^3\Pi_g$, and $A^3\Sigma_u^+$
States of Nitrogen Molecule

State and vibrational level	$\sigma_{max} \cdot 10^{-18}$, cm²				τ, sec	B_v
$C^3\Pi_u$	[74]	[76]	[77]	[73, 75]	[82]	
0	16.12	25.4	23.5	53.8		1.8161
1	9.34	18 6	14.4	30.0		1.7963
2	3.82	7.9	4.9	10.4		1.7767
. . .	. . .		. . .	. . .		. . .
$\sum_v \sigma_{max}$	30.6	51.9	42.8	98.5	$4 \cdot 10^{-8}$	$B_e = 1.8259$
$B^3\Pi_g$	[74]	[80]	[81]	[73, 75]	[82]	
0	11.27	3.7	3.8	3.97		1.6288
1	16.70	7.2	9.2	9.60		1.6104
2	19.36	11.9	11.0	12.70		1.5920
3	16.32	11.8	11.3	12.40		1.5736
4	13.95	10.4	10.3	9.83		1.5552
5	8.14	7.7	7.8	6.85		1.5368
6	5.50	5.4	5.7	4.28		1.5184
7	3.82	3.3	3.6	2.52		1.5000
8	2.68	1.82	1.7	1.40		1.4816
9	2.06	1.03	0.79	0 74		1.4632
10	1.45	0.61	0.41	0.38		1.4448
11	1.10	0.39	0.23	0.19		1.4264
12	0.79	0.29	0.12	0.095		1.4080
. . .				. . .		
	95.7	65.5	65.9	65. 0	$9 \cdot 10^{-6}$	$B_e = 1.6380$
$A^3\Sigma_u^+$				[73]	[83—85]	
0				0.31		1.4335
1				1.65		1.4205
2				4.64		1.4075
3				9.62		1.3945
4				15.2		1.3815
5				21.0		1.3685
6				25.6		1.3555
7				28.7		1.3425
. . .				. . .		. . .
				314	1.3—2.7	$B_e = 1.440$

levels of the laser states. Theoretical and experimental data indicate that the excitation cross sections of vibrational levels are very nearly proportional to the corresponding Franck—Condon factors [73]. Consequently, it is sufficient to know the total excitation cross sections of the electronic states. Such cross sections are calculated for the seven lowest triplet states of the nitrogen molecule, including all the states of interest to us, in [73]. The same paper gives a review of all the experimental and theoretical data on this subject. This review shows that even for such a thoroughly investigated molecule as nitrogen we still do not have full and reliable data on the electron-excitation cross sections. We shall consider the situation in the case of three laser states C, B, and A. We shall not consider the states D and W because their excitation cross sections are relatively small [73, 74] and cascades from these states should not alter significantly the overall level population pattern.

The cross sections σ_{max} corresponding to the maxima of the excitation functions of the electronic states A, B, and C and of the individual vibrational levels (reported by various workers) are collected in Table 5. The sixth column in this table gives the calculated excitation cross sections taken from Cartwright's theoretical paper [73]. He reports only the total cross sections for the states C and B. The cross sections of individual vibrational levels of these states were deduced by us by multiplying the total cross sections taken from [73] by the corresponding Franck—Condon factors taken from [75]. Table 5 gives also the radiative lifetimes and rotational constants. The latter were calculated for the individual vibrational levels from the formula $B_v = B_e - \alpha_e (v + 1/2)$, where the equilibrium values of the constants B_e and α_e were taken from [59].

State $C^3\Pi_u$

The excitation cross sections of this state were investigated experimentally by an optical method in [74, 76–78] and by another method in [79]. The results obtained differed by amounts greater than the estimate errors. The results given in [78, 79] were in agreement with each other and with the calculated values in [73]. The cross sections reported in [76, 77] were about (within 15%) half the values given in [73]. Finally, the cross section given in [74] was approximately half that in [76]. The origin of these discrepancies is not yet clear. At present it is best to adopt the middle values reported in [76, 77].

State $B^3\Pi_g$

The excitation cross sections of this state were determined experimentally (by an optical method in [74, 80, 81]) and theoretically [73]. All the experimental cross sections included a contribution of cascades from higher states. The best agreement between the results of various experimental investigations and the calculations was achieved for this state. However, it was difficult to compare accurately the experimental and calculated values because of the influence of the cascade transitions. The total excitation cross section of the state B was somewhat higher than that of the state C.

State $A^2\Sigma_u^+$

This state is metastable and, therefore, it is difficult to measure its excitation cross section by optical methods. Only theoretical calculations [73] are available and they show that the total excitation cross section of the state A is greater than that of the state B. However, the excitation of the state A is distributed over its many vibrational levels, particularly between the relatively high levels.

These characteristics of the electronic states of the N_2 molecules show that in the 2^+ and 1^+ systems the radiative probability of decay of the lower laser state is considerably less than the probabilities of the active (laser) transitions. This means that under normal conditions, when the decay of various states due to collisions with heavy particles is relatively slow, a steady-state population inversion cannot be achieved for the 2^+ and 1^+ band systems. On the other hand, the same characteristics allow us to identify the mechanism which can ensure pulse inversion of these transitions. In fact, if we assume that the laser levels are populated mainly by direct electron impact from the ground state of the molecule, we find that the relative values of the electron-excitation cross sections of these levels play the dominant role. We have seen that the total excitation cross section of the $C^3\Pi_u$ state is less than that of $B^3\Pi_g$, whereas the total cross section of $B^3\Pi_g$ is, in its own turn, less than that of $A^2\Sigma_u^+$. In this situation a population inversion can be achieved only for some bands in the 2^+ and 1^+ systems as a result of a difference between the populations of various vibrational levels. Thus, the positions of the potential curves and the associated distribution of the Franck–Condon factors over the bands become of decisive importance.

It is clear from Fig. 22 that the positions of the potential curves of the triplet states of the N_2 molecule are such that we can expect a population inversion for some bands in the 2^+ system. An analysis of the Franck–Condon factors in [8] has led to the conclusion that the direct electron excitation should result in a population inversion of the 0–0, 0–1, and 1–0 bands of this system. The above data on the excitation cross sections of the individual vibrational levels of the C and B states lead to the same conclusion. The discrepancy between the results obtained by different workers gives rise to very different estimates of the vibrational inversion coefficients. Among the cited papers the least favorable (for population inversion)

TABLE 6. Vibrational Inversion Coefficient a
for Bands in 2^+ System of N_2
(calculations based on [74])

Band $v' \rightarrow v''$	B'_v / B''_v	N'_v / N''_v	$a = \dfrac{B'_v N'_v}{B''_v N''_v}$
0—0	1.115	1.430	1.60
0—1	1.128	0.965	1.09
1—0	1.103	0.829	0.91

cross sections are reported in [74]. If these results are used in estimating the vibrational inversion coefficients for the 2^+ band system, the values obtained are those listed in Table 6. According to these calculations the highest gain should be obtained for the 0—0 band. Vibrational inversion in the 0—1 band should be considerably less, whereas for the 1—0 band it should be $a < 1$. However, this does not mean that laser emission cannot be obtained in the 1—0 band.

A calculation of the gain given in Chap. I shows that population inversion of some rotational transitions can be achieved also for $a < 1$. At room temperature (~320°K), such inversion is obtained for $a > 0.93$ whereas at liquid nitrogen temperature it occurs for $a > 0.75$. Thus, if we used the least favorable (for population inversion) values of the excitation cross sections, we find that we can expect laser emission for all the bands under consideration, but in the 1—0 band case it should be observed only when the gas is cooled.

The excitation cross sections given in [76, 77] give values of a approximately twice as high as those just quoted. Moreover, these values are obtained using the cross sections of the B state including cascades from the higher states, which are of no importance during the short stimulated radiation pulses considered here. The available information on the characteristics of electronic states and transitions in the nitrogen molecule demonstrates that the direct electron-impact excitation from the ground state of the molecule should give rise to population inversion and laser emission as a result of these transitions in the 2^+ system of nitrogen. It follows from the available data that we cannot exclude the possibility of population inversion in other bands of the 2^+ system, particularly in the 0—2 band.

We can easily see that all the results obtained fit the proposed population inversion mechanism. Laser emission is observed in the 0—0, 0—1, and 1—0 bands. A qualitative distribution of the gain between these bands is in agreement with the results listed in Table 6. The rotational structure of the stimulated emission band is generally in agreement with calculations given in Chap. I. The behavior of laser emission observed when the gas is cooled is also in good agreement with our calculations based on the assumption of direct electron excitation. Finally, the properties of the stimulated radiation are governed by the value of the parameter γ, which is again in agreement with the direct electron-impact mechanism. Moreover, this mechanism is in agreement with the observed short duration of the stimulated radiation pulses, which correspond to the leading edges of the current pulses.

During the investigation reported here several papers appeared reporting results which could be used to identify the population inversion mechanism in the 2^+ nitrogen bands. A calculation of the saturated stimulated radiation power was given in [23-25]. This calculation was made numerically on the assumption of direct electron excitation of the laser levels under discharge conditions used in the experimental investigations reported in [20, 22, 27]. The calculated results were in good agreement with the experiments.

Thus, we may conclude that pulse stimulated emission in bands of the second positive system of the N_2 molecule is due to the direct electron-impact excitation of the active levels from the ground state of the molecule.

CHAPTER IV

LASER EMISSION AND SUPERRADIANCE IN FIRST POSITIVE (INFRARED) SYSTEM OF NITROGEN BANDS

§ 1. Characteristics of Laser Emission from Cooled Active Gas

We investigated the laser emission in the first positive (1^+) system of bands of the N_2 molecule using the same excitation method as in the investigation of the 2^+ system. In all experiments we used "special purity" nitrogen. The main modification of the method was the use of wider discharge tubes since the laser emission power in the 1^+ system was low if narrow tubes were used. The main results reported in the present chapter were obtained using a tube whose internal diameter was 26 mm and which had an external cooling jacket. Some preliminary results were obtained using a tube of 8 mm diameter.

A quartz tube (internal diameter 26 mm and active length 100 cm) had two internal cold Duralumin electrodes located inside the branches. The length of the discharge gap was 135 cm. The capacitance of the working capacitor was 0.01 μF. The duration of the current pulses under optimal (from the point of view of the laser emission power) conditions was 360 nsec at midamplitude and the peak value of the current was close to 2 kA. In most experiments the discharge tube was cooled with liquid nitrogen along the whole of its active length.

A laser emission pulse appeared at the leading edge of an almost triangular current pulse and it reached its maximum at approximately the same time as the current. The duration of the laser emission pulse depended little on the working conditions in the laser and it was ~100 nsec at midamplitude. Laser emission was recorded using an FÉK-09 coaxial photocell with an oxygen-cesium photocathode. The principal parameters of the discharge current were deduced from an oscillogram of the signal across a low-inductance shunt connected in series with the discharge tube. The total inductance of the discharge circuit was estimated to be ~1.5 μH. A resonator was formed by mirrors with multilayer dielectric coatings. The ends of the laser tube, cut at the Brewster angle, were closed by glass windows. The laser emission and superradiance spectra were investigated using a DFS-13 spectrograph in the first order of a grating with 600 lines/mm and a dispersion of 4 Å/mm. The average total stimulated radiation power was measured using a calibrated thermopile with an error of ±20%.

Figure 23 shows two series of the dependences of the total (over the spectrum) energy W of the laser emission pulses on the active gas density N. These dependences were obtained for different voltages V across the working capacitor. One series (Fig. 23a) was recorded without forced cooling (gas temperature $T_g \approx 320°K$) and the other (Fig. 23b) employing liquid nitrogen cooling ($T_g \approx 100°K$). Cooling increased considerably the energy of the laser emission pulses without any additional losses of the energy supplied to the discharge. Moreover, the optimal gas density N_{opt}, corresponding to the maximum energy of the laser emission pulses, and the densities corresponding to the limits of the laser emission region, were found to be functions of V.

Figure 24 shows the experimental values of V and N corresponding to the maximum laser emission energy (denoted by "optimum" in Fig. 24) and the limits of the laser emission region at $T_g = 100°K$. These experimental were found to fit well straight lines converging at

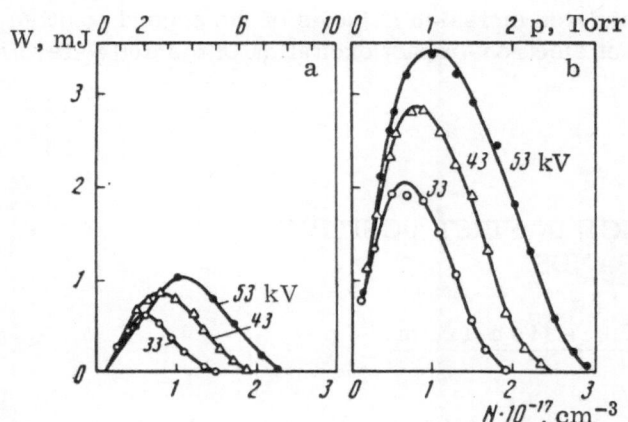

Fig. 23. Dependences of the pulse energy of the laser emission in the 1^+ nitrogen system on the density of the active gas obtained for different voltages across the tube: a) without cooling; b) with liquid nitrogen cooling.

the same place on the ordinate at $V = V_0$ (continuous lines). The outer lines represented the limits of the values of V and N between which laser emission could be obtained. These limits and the optimal conditions could be described by the values of the parameters $\gamma = (V - V_0)/N$, representing the slopes of these lines. The experimental results indicated that γ_{opt} (lines denoted by 2 in Fig. 2) was independent of the gas temperature but the values of γ_{min} and γ_{max} (lines 1 and 3) varied with temperature. The range of conditions suitable for laser emission at 320°K was somewhat narrower than at 100°K. Cooling to 100°K increased the gain so much that high-power superradiance was observed. The peak superradiance power obtained using one mirror under optimal conditions (for a voltage of 53 kV and a pressure of 1.5 Torr) was ~10 kW. Our estimates indicated that the gain was at least 25 dB/m. To the best of our know-

Fig. 24. Range of conditions under which laser emission (continuous lines) and superradiance (dashed lines) were obtained in the 1^+ nitrogen system and the dependences of the optimal conditions on the voltage across the tube V and the active gas density N at T = 100°K: 1) upper limit (γ_{max}); 2) optimum (γ_{opt}); 3) lower limit (γ_{min}).

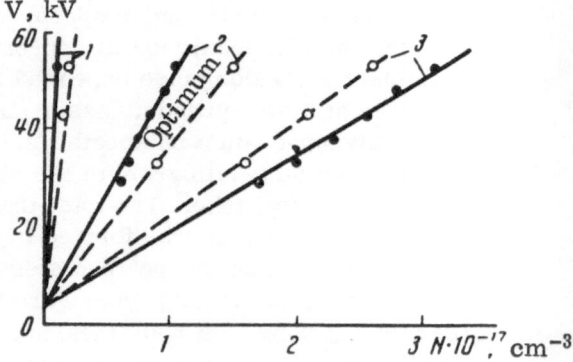

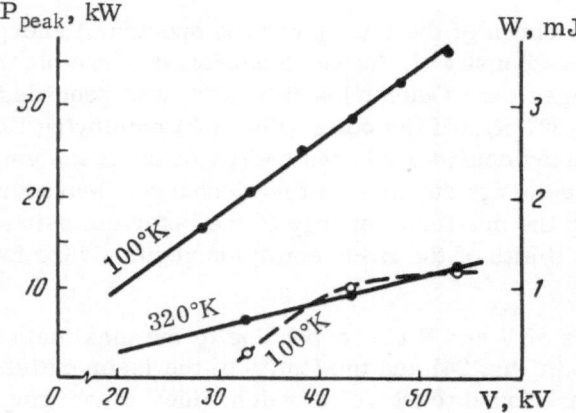

Fig. 25. Dependences of the total peak power of the laser emission (continuous lines) and superradiance (dashed curve) in the 1^+ nitrogen system on the voltage across the tube, plotted for γ_{opt} with and without cooling of the active gas.

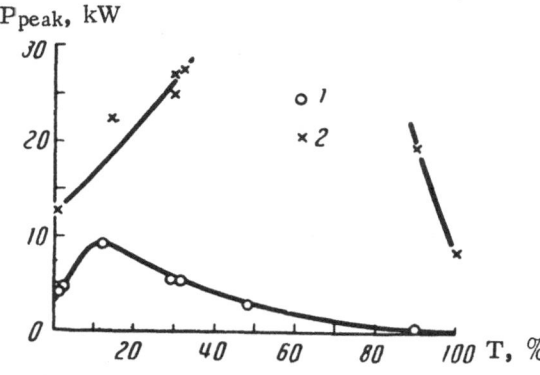

Fig. 26. Dependences of the total peak infrared laser emission power in the 1^+ nitrogen system on the transmission of the exit mirror without cooling, T = 320°K (1), and with liquid nitrogen cooling, T = 100°K (2); V = 43 kV, $\gamma = \gamma_{opt}$.

ledge, this was the second gas-discharge-excited molecular system which exhibited superradiance. Earlier superradiance was observed only for transitions in the ultraviolet part of the spectrum of the 2^+ system (see Chap. III). The range of existence of superradiance at a given temperature was also limited by the lines shown dashed in Fig. 24, which represented different values of the parameter γ. The optimal conditions again corresponded to a fixed γ but its value was different from that for the optimal laser emission conditions.

Figure 25 demonstrates the change in the total peak laser emission power with increasing V for $\gamma = \gamma_{opt}$ with and without cooling of the gas. This figure includes the corresponding curve for superradiance at 100°K. In both cases the power increased continuously when the voltage was raised. The 320°K characteristic was obtained using an exit mirror with a transmission of 13%, which was optimal for achieving the highest output power at this temperature. The 100°K characteristic was obtained using an exit mirror with 30% transmission, which was not optimal. Figure 26 gives the dependences of the total laser emission power on the transmission of the exit mirror obtained with and without forced cooling by liquid nitrogen. In both cases the voltage applied to the capacitor was 43 kV. When the temperature was lowered, i.e., when the gain increased, the optimal transmission of the exit mirror shifted strongly in the direction of higher transparency. Unfortunately, the lack of necessary mirrors prevented us from investigating the transmission range near the maximum for the cooled laser.

In our experiments the maximum total laser emission power corresponded to V = 53 kV, p = 1 Torr, and $T_g = 100$°K when the exit transmission was 30%. This power was 55 kW for laser emission pulses whose duration was 100 nsec at midamplitude. At room temperature the maximum power obtained under optimal conditions (for the same 53 kV) barely reached 10 kW.

The output power of 55 kW was a record for a laser utilizing the 1^+ nitrogen system. The corresponding power density was ~10 kW/cm^2, whereas the total and specific powers inside the resonator were 160 kW and 0.3 kW/cm^3, respectively. We were able to focus the laser beam (using a spherical lens with f = 7 cm) into a spot of ~0.5 mm in diameter. It should be noted that the electric discharge, particularly under conditions corresponding to maximum output power (i.e., at high pressures and voltages), was localized near the tube walls. Consequently, the laser beam and the focused spot were ring-shaped. The power density at the edges of the ring was estimated to be of the order of 100 MW/cm^2.

§ 2. Laser Emission and Superradiance Spectra.

Reversal of Intensity Alternation

Laser emission and superradiance were observed in the following vibrational bands:

$v' \rightarrow v''$ . . .	4—2	3—1	2—0	2—1	1—0	0—0
λ, μ	0.751	0.762	0.776	0.872	0.891	1.05

TABLE 7. Rotational Structure of Infrared Superradiance Bands in
1^+ System of N_2

Band $v' \rightarrow v''$	λ_{meas}, Å (in air)	Rotational transition	Band $v' \rightarrow v''$	λ_{meas}, Å (in air)	Rotational transition
2—1	8704.572	$Q_1 7$	1—0	8904.408	$Q_1 3$
	8710.304	$Q_1 5$		8910.577	$P_{12} 5$
	8715.553	$Q_1 3$	0—0	10490.64	$Q_1 5$
	8721.237	$P_{12} 5$		10498.72	$Q_1 3$
	8722.222	$P_{12} 3$		10507.16	$P_{12} 5$
1—0	8898.918	$Q_1 5$		10508.41	$P_{12} 3$

Note. The rotational transition is identified by the rotational quantum number K″ (ignoring the spin angular momentum) of the lower active state; the branch Q_1 corresponds to transitions satisfying the conditions $\Delta K = K' - K'' = 0$ and $\Delta J = J' - J'' = 0$; the branch P_{12} corresponds to transitions satisfying the conditions $\Delta K = -1$ and $\Delta J = 0$.

Superradiance was observed only when a mirror was present. The superradiance power obtained from a cooled laser was concentrated largely in the 2—1 and 1—0 infrared bands. The red superradiance bands 4—2, 3—1, and 2—0 were very weak and they appeared only under very favorable conditions. Information on the rotational structure of the strongest (infrared) superradiance bands is collected in Table 7. Relatively few lines appeared in the superradiance spectrum: They were the lines of the Q_1 and P_{12} branches belonging to the ortho-modification of nitrogen.

In contrast to superradiance, the distribution of the laser emission power between the bands depended strongly on the working conditions and particularly on the selected mirrors. Under conditions corresponding to the maximum value of the peak power the laser mirrors concentrated a considerable proportion of the power in the red bands. At the power levels obtained, these bands were not only clearly visible with the naked eye but they also gave rise to a bright red spot on a white sheet of paper. The infrared bands were also readily visible to the naked eye when passed through an infrared filter; they appeared as white light.

The distribution of the laser emission power within the bands themselves also depended on the conditions inside the laser tube. For example, when the gas was cooled, the changes which occurred in the band spectrum were of dual nature: We observed a redistribution of the output power between the vibrational branches of the band and some increase in the number of lines; we also found new rotational branches. At room temperature the laser emission was concentrated mainly in the Q_1 and P_{12} branches (the Q_1 branch was much stronger) with the maximum power in the lines for which the rotational quantum number of the lower level was K″ = 9; at low temperatures, we found additional rotational branches Q_3, P_{23}, and O_{12} with a power maximum shifted in all of them in the direction of lower values of K. Moreover, the intensities of the side branches P_{12}, P_{23}, and O_{12} increased and became comparable with the intensity of the main branch Q_1. The nature of the stimulated emission spectrum in the region of the 2—1 and 1—0 bands emitted from the cooled gas is shown schematically in Fig. 27.

The laser emission spectrum in the region of the 1^+ nitrogen bands was very rich in lines. It had been investigated quite thoroughly before [17, 52, 57]. Our laser emission spectrum was similar, apart from the features associated with the influence of cooling. We observed a few new lines. Since the majority of the laser emission lines was already measured and attributed to specific transitions, and the new lines did not alter the pattern very significantly, we did not carry out detailed measurements of the laser emission spectra of all the bands. We concentrated our attention on the 4—2, 3—1, and 2—0 bands for which only a few lines had been reported earlier. The results of our measurements are given in Table 8.

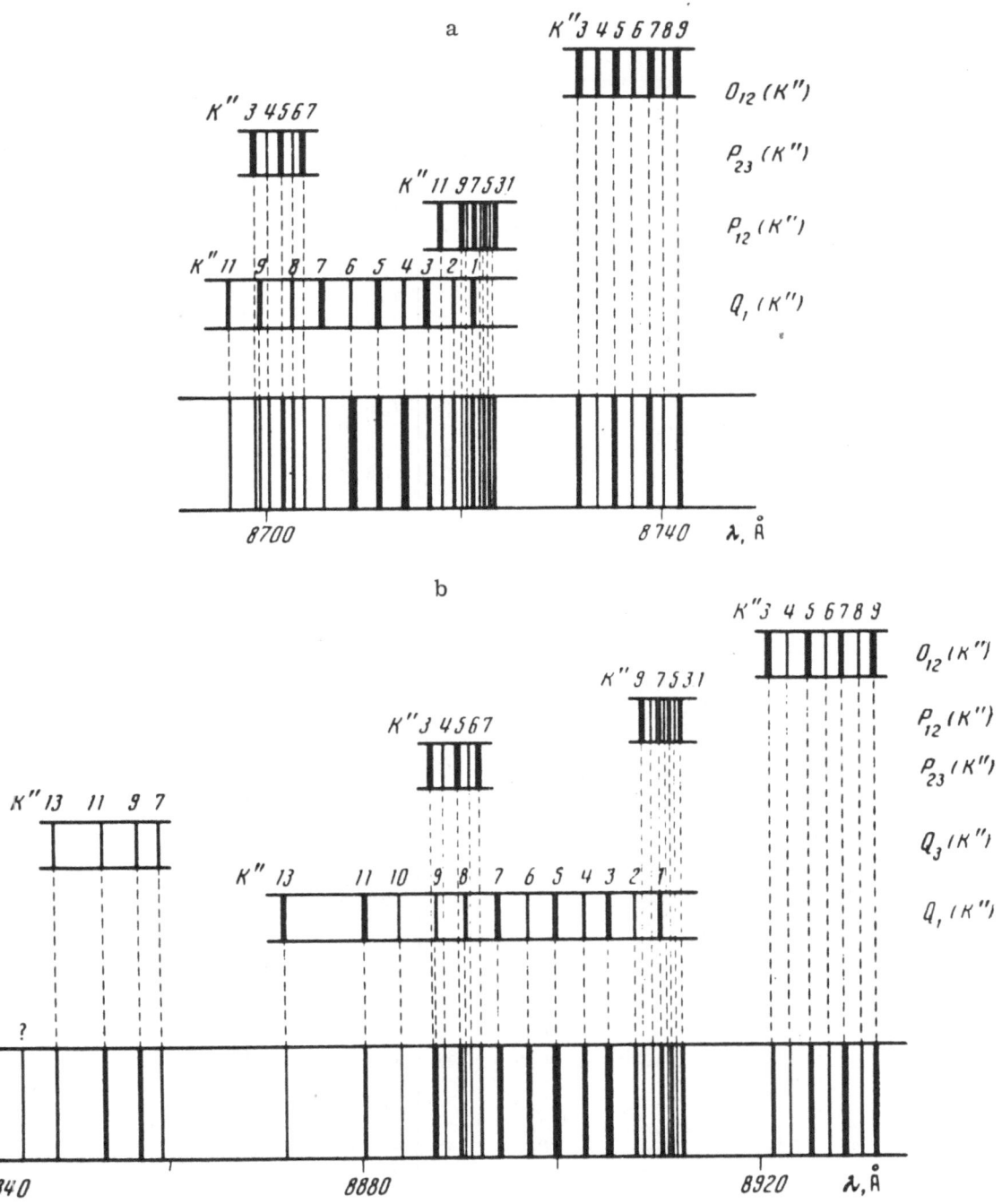

Fig. 27. Schematic representation of the spectra of the 2−1 (a) and 1−0 (b) laser emission bands in the 1$^+$ nitrogen system obtained using a tube cooled with liquid nitrogen, showing identification of rotational transitions. Here, K″ is the rotational quantum number (without the spin momentum) of the lower active state for Hund's coupling case b.

TABLE 8. Rotational Structure of Laser Emission Bands in 1^{+} System of N_2 at $T_g = 100°K$

No.	λ_{meas}, Å (in air)	Rotational transition		No.	λ_{meas}, Å (in air)	Rotational transition	
		2−0 band					
1	7773.15 *	$O_{12}14$	(?) $^{o}P_{12}14$	14	7744.46 *	—	—
2	7765.76 *	$O_{12}7$	$^{o}P_{12}7$	15	7743.88	$Q_{1}5$	$Q_{1}6$
3	7763.99 *	$O_{12}6$	$^{o}P_{12}6$	16	7743.28 *	—	—
4	7762.81 *	$O_{12}5$	$^{o}P_{12}5$	17	7742.60 *	—	—
5	7759.62 *	$O_{12}3$	$^{o}P_{12}3$	18	7742.27 *	—	—
6	7753.74 *	$P_{13}5$	$^{P}R_{13}4$	19	7742.04 *	—	—
7	7752.98 *	$P_{13}7$	$^{P}R_{13}6$	20	7741.80 *	$Q_{1}6$	$Q_{1}7$
		$P_{12}3$	$^{P}Q_{12}3$	21	7739.65	$Q_{1}7$	$Q_{1}8$
8	7752.45 *	$P_{12}5$	$^{P}Q_{12}5$	22	7738.03 *	$P_{23}7$	$^{P}Q_{23}6$
9	7747.81 *	$Q_{1}3$	$Q_{1}4$	23	7737.50 *	$Q_{1}8$	$Q_{1}9$
10	7747.20 *	—	—	24	7736.28 *	$P_{23}5$	$^{P}Q_{23}4$
11	7745.88 *	$Q_{1}4$	$Q_{1}5$	25	7735.02	$Q_{1}9$	$Q_{1}10$
12	7745.24 *	—	—	26	7733.92 *	$P_{23}3$	$^{P}Q_{23}2$
13	7744.96 *	—	—	27	7733.54 *	—	—
		3−1 band					
1	7640.81 *	$O_{12}9$	$^{o}P_{12}9$	13	7622.23 *	$Q_{23}7$	$^{o}P_{23}6$
2	7638.27 *	$O_{12}7$	$^{o}P_{12}7$	14	7621.18 *	$Q_{1}3$	$Q_{1}4$
3	7636.84 *	$O_{12}6$	$^{o}P_{12}6$	15	7619.34 *	$Q_{1}4$	$Q_{1}5$
4	7635.45 *	$O_{12}5$	$^{o}P_{12}5$	16	7617.36	$Q_{1}5$	$Q_{1}6$
5	7633.97 *	$O_{12}4$	$^{o}P_{12}4$	17	7615.33 *	$Q_{1}6$	$Q_{1}7$
6	7632.45 *	$O_{12}3$	$^{o}P_{12}3$	18	7613.26	$Q_{1}7$	$Q_{1}8$
7	7626.73 *	$P_{13}5$	$^{P}R_{13}4$	19	7611.51	$P_{23}7$	$^{P}Q_{23}6$
8	7626.01 *	$P_{13}7$	$^{P}R_{13}6$	20	7611.04 *	$Q_{1}8$	$Q_{1}9$
9	7625.66	$P_{12}5$	$^{P}Q_{12}5$	21	7609.84 *	$P_{23}5$	$^{P}Q_{23}4$
10	7625.14	$P_{12}7$	$^{P}Q_{12}7$	22	7608.78	$Q_{1}9$	$Q_{1}10$
11	7624.65 *	$Q_{1}1$	$Q_{1}2$	23	7607.62 *	$P_{23}3$	$^{P}Q_{23}2$
12	7622.92	$Q_{1}2$	$Q_{1}3$	24	7581.03 *	—	—
		4−2 band					
1	7525.64 *	—	—	23	7499.36 *	$Q_{1}3$	$Q_{1}4$
2	7522.94 *	—	—	24	7498.63 *	$Q_{12}3$	$^{?Q}R_{12}3$
3	7520.07 *	—	—	25	7497.52 *	$Q_{1}4$	$Q_{1}5$
4	7517.89 *	—	—	26	7496.86 *	$Q_{12}4$	$^{?Q}R_{12}4$
5	7516.90 *	$O_{12}8$	$^{o}P_{12}8$	27	7496.65	—	—
6	7516.59	—	—	28	7495.67	$Q_{1}5$	$Q_{1}6$
7	7515.58 *	$O_{12}7$	$^{o}P_{12}7$	29	7495.00 *	$Q_{12}5?$	$^{Q}R_{12}5$
8	7514.29 *	$O_{12}6$	$^{o}P_{12}6$	30	7493.67 *	$Q_{1}6$	$Q_{1}7$
9	7513.96 *	—	—	31	7492.96 *	$Q_{12}6?$	$^{Q}R_{12}6$
10	7513.03 *	$O_{12}5$	$^{o}P_{12}5$	32	7492.69 *	—	—
11	7511.53 *	$O_{12}4$	$^{o}P_{12}4$	33	7491.72	$Q_{1}7$	$Q_{1}8$
12	7510.08 *	$O_{12}3$	$^{o}P_{13}3$	34	7491.00 *	$Q_{12}7?$	$^{Q}R_{12}7$
13	7509.21 *	—	—	35	7489.77 *	$Q_{1}8$	$Q_{1}9$
14	7505.63 *	—	—	36	7489.29 *	$P_{23}6$	$^{P}Q_{23}5$
15	7504.56 *	$P_{13}5$	$^{P}R_{13}4$	37	7488.24 *	$P_{23}5$	$^{P}Q_{23}4$
16	7503.98 *	$P_{13}7$	$^{P}R_{13}6$	38	7487.40	$Q_{1}9$	$Q_{1}10$
17	7503.69 *	$P_{12}5$	$^{P}Q_{12}5$		*	$P_{23}4$	$^{P}Q_{23}3$
18	7503.02 *	$P_{12}7$	$^{P}Q_{12}7$	39	7486.13 *	$P_{23}3$	$^{P}Q_{23}2$
19	7502.69 *	$Q_{1}1$	$Q_{1}2$	40	7485.62 *	—	—
20	7501.02 *	$Q_{1}2$	$Q_{1}3$	41	7483.70 *	—	—
21	7500.52 *	—	—	42	7481.69 *	—	—
22	7500.30 *	—	—	43	7478.26 *	—	—

TABLE 8. Continued

No.	λ_{meas}, Å (in air)	Rotational transition		No.	λ_{meas}, Å (in air)	Rotational transition	
				4—2 band			
44	7476.15 *	—	—	49	7472,95 *	—	—
45	7475.62 *	—	—	50	7471,68 *	—	—
46	7475.21 *	—	—	51	7471,17 *	—	—
47	7473.89 *	—	—	52	7469,31 *	—	—
48	7473,61 *	—	—	53	7465,62 *	—	—

Note. Two notation systems are used to identify the rotational transition: the system on the left is based on the selection rules obeyed by the rotational quantum number K (giving the value of K" for the lower active state), which is more suitable for the present case; the system on the right is used more widely but it is less suitable for the present case and it is based on the selection rules obeyed by the rotational quantum number J (the value of J" is given for the lower state). The symbol (?) is used for the transitions which are identified only tentatively. The asterisks denote new laser emission lines.

The new laser emission lines are identified by asterisks. (When the present paper was being prepared for press, we became acquainted with an investigation [56] for the laser emission spectrum of the 1^+ nitrogen system. The experiments were carried out under conditions similar to ours and many of the lines listed in Table 8 were observed.)

We shall consider in somewhat greater detail one particular feature of the laser emission spectra observed by us at low temperatures. It is known [59] that the spectra of symmetric molecules are characterized by an alternation of intensities in the rotational-vibrational bands associated with the influence of the nuclear spin. This alternation is observed easily in the spontaneous and stimulated radiation spectra of the 1^+ nitrogen bands. However, when the gas was cooled we found, unexpectedly, that under certain conditions some of the rotational branches exhibited a reversed alternation of intensities. Figure 28 shows two laser emission spectrograms of the 2—1 band obtained for a gas cooled with liquid nitrogen under identical conditions (the only difference was in the resonator mirrors). The alternation of intensities is clearly visible in these spectrograms. However, it is worth noting that the alternation of intensities in the Q_1 branch is opposite in the two cases. The spectrogram b shows the normal alternation: the lines with the odd values of K" are stronger. This alternation is reversed in the spectrogram a: here, the stronger lines are those with even values of K". Such a reversal of the alternation of intensities in the rotational spectra has not yet been reported, to the best of our knowledge, for spontaneous or stimulated radiation.

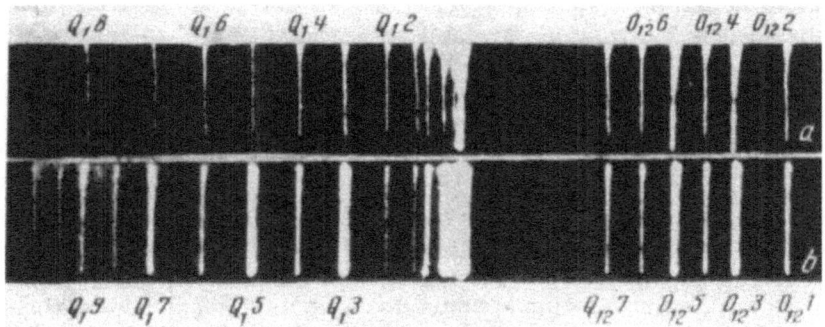

Fig. 28. Spectrograms of the 2—1 laser emission band in the 1^+ nitrogen system: a) inversion of the alternation of intensities in the Q_1 branch; b) normal alternation of intensities in the Q_1 branch.

This reversal was observed only for the Q_1 branches of the 2−1 and 1−0 bands and only under certain experimental conditions. Usually the laser emission spectrum had the normal form. The reversal was most frequently observed near the maximum of the rotational distribution closer to small values of K (lines with K″ = 2-6). It was interesting that the alternation of intensities in the O_{12} branch remained the normal. It was also noted that a suitable selection of mirrors could give both normal and reversed alternation of intensities.

§3. Discussion of Results

The experimental results reported above demonstrated that, as in the case of the 2^+ system, the characteristic of the laser emission in the 1^+ system were governed, at any temperature, by the gas density and the voltage applied to the working capacitor, or, more exactly, by the parameter $\gamma = (V - V_0)/N$. The maximum laser emission power was always reached at some constant (for a given tube) value γ_{opt} and the laser emission stopped at γ_{max} and γ_{min}. Some difference between the values of the latter two parameters between 320 and 100°K could be attributed to the existence of a threshold and a lower gain at 320°K. The parameter γ governed also the properties of superradiance. Slight differences between the optimal conditions (γ_{opt}) for laser emission and superradiance could be explained, for example, by the observation that the main contribution to the total power was made by different bands characterized by different optimal conditions.

Thus, in this case too the population inversion was due to the interaction of molecules with electrons and its magnitude was governed by the average energy of electrons in a discharge. The meaning of the parameter γ was discussed in greater detail in the preceding chapter.

An important factor to note was that the laser emission power corresponding to $\gamma = \gamma_{opt}$ increased approximately linearly with the voltage. This indicated that a considerably higher peak power and energy of the laser emission could be expected by the use of higher gas densities and voltages. If the main processes responsible for the population inversion should remain unchanged up to gas densities corresponding to the atmospheric pressure, extrapolation of the data obtained so far suggested peak powers of the order of several megawatts from the system under consideration.

Our investigation of the influence of the gas temperature on the laser emission in the 1^+ nitrogen system demonstrated that, as in the case of the 2^+ system, cooling resulted in a considerable increase of the laser emission power (for the same value of the gas density and voltage). The redistribution of the power in the rotational structure in the direction of lower values of J, change in the range of conditions under which laser emission was obtained, and other results supported the conclusion that the increase in the power due to cooling resulted from the increase in the gain. A direct proof of a considerable increase in the gain was provided by the observation of low-temperature superradiance.

Since the excitation (current) pulses applied at a given gas density were practically independent of temperature, we assumed that the total populations of the active vibrational levels were not affected by cooling but simply a redistribution took place between the rotational sublevels. Calculations in Chap. I indicated that, in combination with a reduction in the line width, this was responsible for the considerable increase in the gain. Thus, the increase in the laser emission power and energy as a result of cooling of the active gas should be attributed not to a change in the pumping conditions but to a fuller utilization of the population inversion of the active levels. This fuller utilization became possible as a result of an increase in the gain, i.e., because of the more rapid rise of the electromagnetic field in the resonator. Thus, the laser emission in the 1^+ system of the N_2 molecule demonstrated clearly the role of the gain and rate of rise of the optical field in pulse lasers and particularly those utilizing molecular transitions.

Cooling with liquid nitrogen, keeping the excitation pulses constant, increased the peak power and energy of the laser emission pulses by a factor of 5.5. Since there was no change in the energy supplied to excite the gas, we concluded that the laser efficiency also increased by a factor of 5.5. When the gas density was optimal and the voltage across the working capacitor was 53 kV, the laser emission efficiency, defined as the ratio of the energy of the laser emission pulses to the energy supplied to the discharge during a laser emission pulse, was ~0.1%. The total peak power then reached 55 kW (0.3 kW/cm^3 inside the resonator). The practical efficiency of the laser, i.e., the ratio of the laser emission energy to the total energy supplied in the form of the excitation (current) pulses, was then 0.03%. Fuller optimization of the mirrors should increase the power and efficiency still further. We found (Fig. 26) that optimization of the mirrors could double the power and efficiency.

The quoted values of the power and efficiency were obtained under conditions corresponding to the highest value of the peak power. Since the laser emission power was related linearly to the voltage across the working capacitor (Fig. 25) and quadratically to the energy stored in this capacitor, the laser efficiency was higher at lower voltages. For example, at 30 kV the efficiency was approximately twice as high (0.2%) as at 55 kV, but the power was only half that at the higher voltage.

Theoretical estimates of the ultimate values of the specific peak power and laser efficiency utilizing the 1$^+$ system were given in [57] for the total saturation case: the values predicted were 1.5 kW $\cdot$ cm^{-3} $\cdot$ Torr^{-1} and 1%, respectively. The experimental values were 0.15 kW $\cdot$ cm^{-3} $\cdot$ Torr^{-1} and 0.1%, which were considerably lower than the theoretical estimates. A more thorough optimization of the resonator and improvements in the homogeneity of the active medium, particularly in the transverse direction, should result in a closer approach to these theoretical values.

An increase in the power and efficiency of the laser utilizing the 1$^+$ nitrogen system was achieved by cooling the gas, which was not necessarily a convenient practical measure. However, the results demonstrated that the population inversion was strong and that any methods which increased the gain would increase also the power and efficiency. The use of higher active gas densities would be of greatest interest. In this case, for a given electron density and energy, pumping of the upper level should increase proportionally to the gas density and this would result in a corresponding increase in the population inversion and gain. At a sufficiently high gas density the gain should be sufficient to ensure the necessary rate of rise of the field also at room temperature. In this case the specific power and efficiency of the laser should approach the ultimate theoretical values. However, one should remember that the optimum conditions corresponded to a constant value of γ. Therefore, an increase in the density in the ordinary longitudinal-discharge tubes would require a corresponding increase in the working voltage. It would then be desirable to consider sectional tubes or transverse-discharge systems and to investigate laser emission under these conditions. The results reported in the present paper suggested that this should make it possible to increase considerably the specific power and to reach efficiencies of the order of 1%.

The laser emission and superradiance spectra of the 1$^+$ system were basically similar to those reported in [17, 52, 56, 57]. The nature of these spectra and the influence of temperature were in good agreement with the ideas put forward in Chap. I. The distribution of the gain and laser emission power between the rotational lines was in agreement with the nature of the Hönl—London factors for the transitions under investigation. The distribution of the gain between the various bands was described well by the corresponding Franck—Condon factors.

The reversal of the alternation of intensity, observed at low temperatures, should be considered separately. Judging by the available data, it was due to the competition between the laser emission lines. In principle, the competition could be of two kinds: between the various bands with common upper or lower active vibrational levels, and between various vibrational branches within one band. We shall consider these two kinds of competition separately.

In the system under discussion, the 2−1 and 1−2 bands, for which the reversal of the intensity alternation was observed, may be in competition with the red bands 3−1 and 2−0, sharing the lower levels with the two former bands. The intensity of the red bands rises more rapidly with falling temperature than does that of the infrared bands, and at liquid nitrogen temperature they become comparable. A suitable selection of the resonator mirrors may be used to vary within wide limits the relative intensities of the red and infrared laser emission bands, and one can envisage conditions under which the red bands are stronger than the infrared bands or vice versa.

Within one band the branch Q_1 may be in competition with the side branch O_{12}, which increases strongly as a result of cooling of the gas and becomes comparable in intensity with the Q_1 branch. However, in view of the proximity of these bands an interchange of the mirrors does not affect their relative intensities. It follows from our experiments that the replacement of the mirrors resulting in the attenuation of the red bands causes a transition from the reversed alternation of the intensities to the normal form. This means that this reversal is due to the competition between the red and infrared bands.

The results reported above, taken as a whole, allow us to draw certain conclusions on the population inversion mechanism. There are two points of view on the mechanism of inversion in the 1^+ system. It is suggested in [1] that the upper active state is populated by a multistage process: First, direct electron impact excites the singlet state $a^1\Pi_g$ from the ground state of the molecule and then collisions of the second kind transfer this excitation to the upper active state $B^3\Pi_g$. A different view is put forward in [8, 52, 57]: in this case it is assumed that the population inversion in the 1^+ system is due to the direct excitation of the active levels by electrons from the ground state of the molecule. In the latter case the changes in the laser emission as a result of cooling of the gas should be generally analogous to those observed for the 2^+ system. This is confirmed by our experiments.

Since the difference between the energies of the levels in the $a^1\Pi_g$ state, populated effectively by electron impact from the ground state, and the levels of the state $B^3\Pi_g$, from which laser emission begins, is fairly large, we may expect the rate of excitation by collisions of the second kind to fall rapidly when the temperature is lowered. Therefore, in the case of excitation by collisions of the second kind, the laser emission power of the 1^+ nitrogen system should fall as a result of cooling, as in the case of the helium−neon laser [86-87]. The reverse is observed experimentally. Thus, the cooling experiments confirm the direct electron-impact mechanism and are not in agreement with the hypothesis of excitation by collisions of the second kind via the $a^1\Pi_g$ state.

At present, mainly due to the work of Knyazev [57], we can regard it as reliably established that the principal population inversion mechanism of the transitions in the 1^+ nitrogen system is the direct excitation of the active levels by electrons from the ground state of the molecule. The results given in the present chapter (see also [88, 89]) provide an additional independent argument in support of this mechanism.

CHAPTER V

LASER EMISSION IN ÅNGSTROM (VISIBLE) BAND SYSTEM OF CARBON MONOXIDE

§ 1.   Characteristics of Laser Emission.

Influence of Cooling

Our experiments on CO were carried out using the apparatus described before. We employed two quartz tubes with internal cold thoriated tungsten electrodes placed inside the branches. The internal diameter of the narrow tube was 8 mm and its discharge length was 111 cm; the internal diameter of the wide tube was 26 mm and its discharge length was 130 cm. The voltage applied to the working capacitor ranged from 8 to 60 kV. The peak current was 0.1-1.8 kA, depending on the experimental conditions. The pulse repetition frequency was 2 Hz. The parameters of the discharge current pulses were determined from oscillograms of the voltage pulses across a low-inductance coaxial shunt connected in series with the discharge tube. These parameters were also determined using a Rogowski loop, operated as a current transformer. The two methods gave identical results.

The laser emission and spontaneous radiation spectra were investigated photographically and oscillographically. In the photographic method we used a DFS-13 spectrograph with diffraction gratings of 1200 and 1600 lines/mm. A DFS-12 spectrograph and a modified form of DFS-13 were used in the determination of the time characteristics of the output radiation. We recorded the total radiation emitted in the form of the two or three strongest rotational lines in the investigated band. We used FÉU-28 and FÉU-36 photomultipliers in the visible range, and FÉU-39 in ultraviolet wavelengths. The photocultiplier signals were recorded using an S1-11 oscillograph with a time resolution of about 10 nsec. Oscillograms of the current and radiation pulses were obtained simultaneously using an S1-17 double-beam oscillograph with a time resolution of 40 nsec. The total (of all the lines) and average values of the laser emission power were measured using a calibrated thermopile and M-95 and M-17/13 galvanometers.

We used CO prepared by the chemical group at the Optical Laboratory of the Lebedev Physics Institute by decomposition of formic acid, as well as ready-made gas from an industrial cylinder. No additional purification was carried out. Carbon monoxide prepared from formic acid usually contained nitrogen as an impurity. The monoxide taken from an industrial cylinder also probably contained this impurity; the other impurities (for which no data were available) did not exceed 1%. The results of the experiments were practically independent of the source of CO. All the measurements, with the exception of one specially discussed case, were carried out on a static gas.

Preliminary experiments demonstrated that the laser emission power in the Ångstrom bands of CO increased strongly when the gas was cooled. Therefore, most of the measurements of the laser emission characteristics were carried out at low temperatures. Room-temperature laser emission was so weak that it was difficult to measure.

Investigations with Narrow Tube [90]

A considerable proportion of the measurements of the laser emission characteristics was carried out using a discharge tube whose internal diameter was 8 mm and which was cooled with liquid nitrogen. Some difficulties were encountered when the temperature of CO was lowered. It was found that a discharge in CO in a narrow tube resulted in a strong fall of the gas pressure below 125°K. Cooling with liquid nitrogen reduced the gas pressure from several torr to a value of the order of 10^{-2} Torr in several minutes from the beginning of a discharge. The addition of a new portion of the gas practically restored the stimulated emission conditions but the need to regulate the pressure resulted in a considerable scatter of the measured values. Therefore, most of the measurements were carried out at a temperature exceeding 125°K. However, it was established that cooling to liquid nitrogen temperature caused a further rise of the gain and laser emission power.

Since the nature of the dependences of the laser emission characteristics on the voltage across the tube, current, and gas pressure were basically the same for all the investigated bands, we confined ourselves to a detailed analysis of the properties of the 0−4 band, which was emitted most easily.

The optimal conditions were determined by the photographic photometry method measuring the dependences of the average laser emission power at the maximum of a rotational distribution (i.e., for J_{max}) on the voltage applied to the tube containing the active gas at different pressures. These measurements were carried out for the following temperatures of the outer wall of the discharge tube: 296, 265, 232, 195, 160, and 125°K. The dependences obtained at 232°K are shown in Fig. 29 (similar dependences were obtained at the other temperatures). It is clear from Fig. 29 that the average laser emission power did not rise mono-

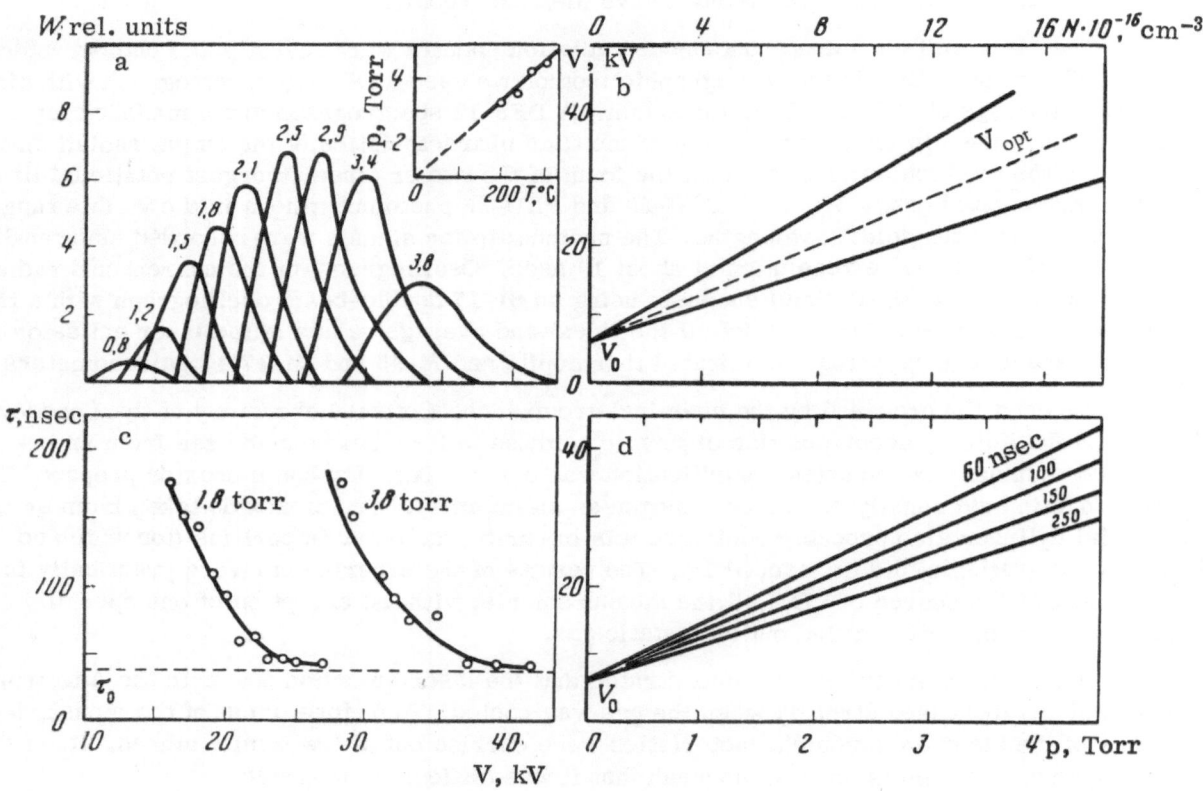

Fig. 29. Power and time characteristics of laser emission in the 0−4 band of the Ångstrom system of CO at T = 232°K: a) dependences of the energy of the laser emission pulses on the voltage obtained at various gas pressures (given in Torrs alongside the curves); b) dependences of the range of existence of laser emission on the parameters V and p (or N), where the upper curve shows V_{max} and the lower curve V_{min}; c) dependences of the duration of laser emission on the voltage obtained at various gas pressures (τ_0 is the radiative lifetime of the upper active level); d) laser emission isochrones plotted in ordinates (V, p or N).

tonically with the applied voltage V but reached a maximum at V_{opt} and then fell rapidly to zero. On the other hand, both V_{opt} and the range of existence of laser emission depended on the gas pressure: the higher the pressure, the greater the shift in the direction of higher voltages. We plotted the dependences of V_{min}, V_{opt}, and V_{max} on the gas pressure (Fig. 29b) and we found that, as in the case of laser emission of the nitrogen bands, extrapolation of these linear dependences toward lower pressures resulted in intersection at one point. This point was located on the voltage axis at $V = V_0$. Similar dependences were obtained also at the other temperatures and at all temperatures the value of V_0 was the same: Only the slope of the lines increased with the gas temperature.

An investigation of the time characteristics of the laser emission revealed that the duration of the laser emission pulses was not constant, as assumed so far, but depended strongly on the applied voltage and gas pressure. It is clear from Fig. 29c that at a constant temperature the duration of stimulated emission pulses τ decreased monotonically when the voltage or pressure was increased. In this way, the duration could be reduced from ~300 to ~30 nsec. (in the absence of reabsorption of the radiation in the ground state, the radiative lifetime of the upper laser level was $\tau_0 = 20$ nsec.) This explicit dependence of the duration of the laser emission pulses on the working conditions could explain the contradictions between the reported values of the duration [2, 11-13, 58]. It was interesting that plotting of equal-duration lines (isochrones) in the coordinates (V, p) yielded straight lines of different slopes but converging at the same point on the ordinate at $V = V_0$ (Fig. 29d). It was remarkable that the value of V_0 found by extrapolation of the time characteristics was (within the limits of the experimental error) the same as that deduced from the power characteristics. The slope of the isochrones increased with decreasing τ. Similar measurements at the other temperatures gave analogous results.

We also found (Fig. 29a) that at a given gas temperature there was some optimal pressure (about 2.7 Torr) at which the highest average laser emission power was obtained for V_{opt}. Experiments showed that the temperature dependence of the optimal pressure (top part of Fig. 29a) could be approximated satisfactorily by a straight line passing through the origin. Since in these experiments the system contained a large volume of the gas which was connected to the tube but was not cooled, the gas pressure in the working part of the tube was practically unaffected by cooling. Then, the gas pressure, density, and temperature were related by $p = NkT$. This relationship and Fig. 29a demonstrated that throughout the investigated temperature range the optimal laser emission $N_{opt} = 1.2 \cdot 10^{17}$ cm^{-3}.

When the gas pressure was converted to the density (Figs. 29b and 29d), it was found that the dependences for V_{min}, V_{opt}, V_{max}, and τ plotted in the coordinates (V, N) were almost identical at all temperatures. This is demonstrated in Fig. 30, which gives an isochrone for $\tau = 100$ nsec. Clearly, systematic variations with temperature are of the same order as the scatter of the points.

Since the straight lines in Figs. 29b and 29d intersected at the same point, an attempt was made to describe the characteristics of laser emission by dependences on the parameter γ, as was done earlier for nitrogen. Figure 31 gives the dependences of the average laser emission power, reduced to the power at γ_{opt}, and laser pulse duration on the parameter γ. We found that, within the limits of the experimental error, these dependences were universal and satisfied all our experimental results.

Figure 32 shows the dependence of W_{max}, corresponding to γ_{opt}, on the voltage. We can see that W_{max} reached its highest value at 26 kV and then fell rapidly. In the range $V > 40$ kV, laser emission from the narrow tube was not observed under any conditions.

As pointed out earlier, all the dependences considered here were obtained for the strongest laser emission in the 0−4 band region. Similar behavior was observed for the other laser

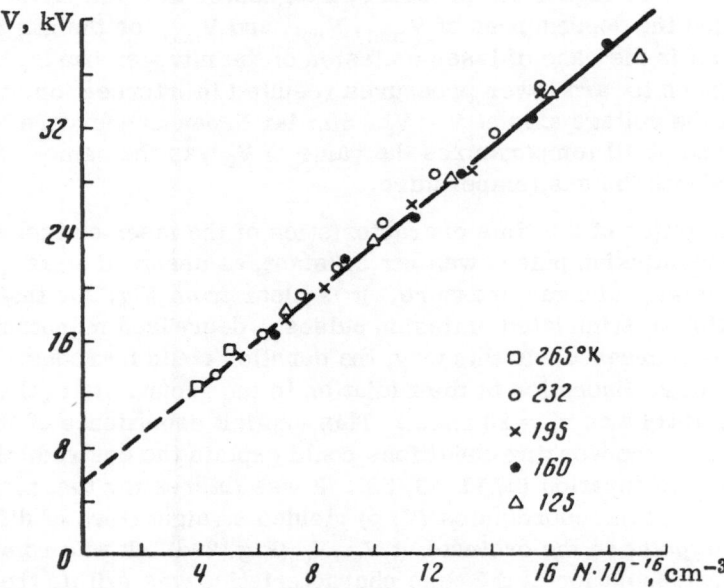

Fig. 30. Scatter of the experimental points in measurements of the duration of laser emission in the 0−4 band of the Ångstrom system of CO at various gas temperatures.

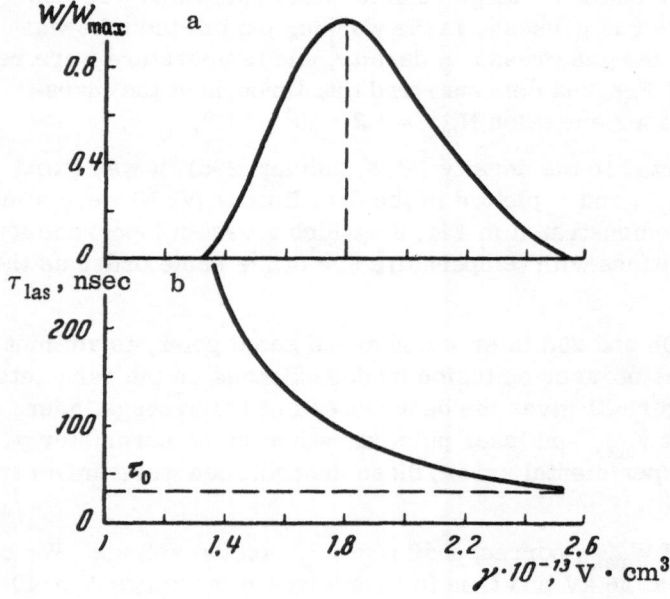

Fig. 31. Dependences of the distribution of the energy (a) and duration (b) of laser emission in the 0-4 band of the Ångstrom system of CO on the parameter γ (for a tube of 8 mm diameter).

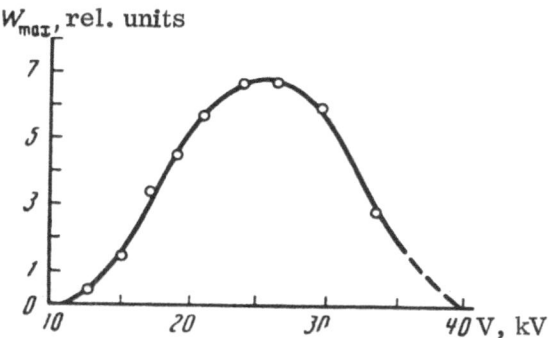

Fig. 32. Dependence of the peak laser emission power in the 0−4 band of the Ångstrom system of CO on the voltage applied to a tube 8 mm in diameter $(\gamma = \gamma_{opt})$.

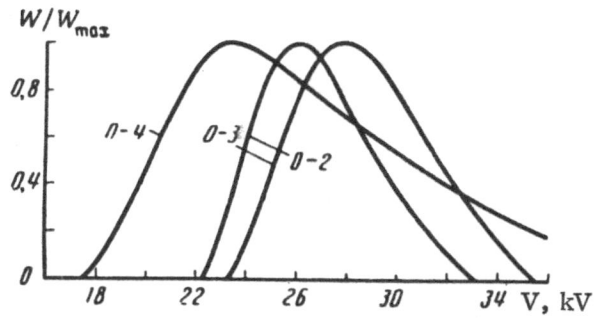

Fig. 33. Dependences of the energy of the laser emission pulses corresponding to the 0−2, 0−3, and 0−4 bands of the Ångstrom system of CO on the voltage applied to the tube $(N_0 = 1 \cdot 10^{17}$ cm^{-3} and T = 180°K).

emission bands in the Ångstrom system. However, the conditions optimal for laser emission differed considerably from band to band. By way of example, we plotted in Fig. 33 the dependences of the average laser emission power in the 0−2, 0−3, and 0−4 bands on the voltage, obtained at the same gas density and temperature. It is clear from this figure that the highest power in all three bands was obtained at different values of V_{opt}. The differences were so great that one could observe visually changes in the color of the stimulated emission (for example, from red to green) when the applied voltage (or gas pressure) was varied. This feature of the laser emission in the Ångstrom system of CO made it possible to tune the laser emission between the bands by altering the voltage across the working capacitor or by changing the gas pressure.

Figure 34 shows oscillograms of the laser emission and current pulses obtained using the S1-17 double-beam oscillograph. Laser emission was observed at the leading edge of the current pulse during its high−current phase. This emission was delayed by about 150 nsec relative to the beginning of the high−current phase. (We included in this delay the time taken by the signal to reach the oscillograph.) An oscillographic study of the spontaneous radiation

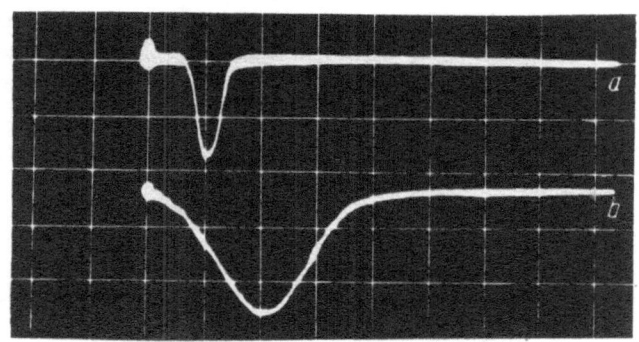

Fig. 34. Oscillograms of the stimulated emission of the 0−4 band in the Ångstrom system of CO (a) and current (b) pulses (for a tube 8 mm in diameter). The horizontal scale is 0.5 μsec/div.

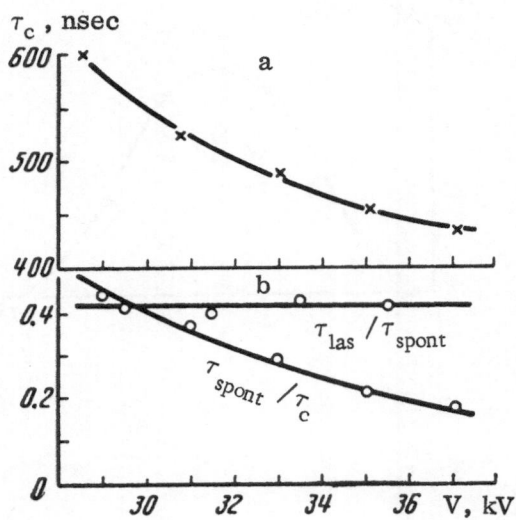

Fig. 35. Voltage dependences of the duration
of the current pulses (a) and of the ratios of
the durations of the current, laser emission,
and spontaneous radiation pulses (b) corresponding to the 0−4 band in the Ångstrom system of CO (for a tube 8 mm in diameter).

(emitted from the end of the discharge tube) demonstrated that the radiation generated as a
result of transitions from either of the active levels was considerably shorter than the current
pulses, and the radiation originating from the upper level reached its maximum slightly earlier
than that originating from the lower level. Laser emission was emitted slightly ahead of the
spontaneous radiation from the upper level. When the voltage was increased and the gas pressure reduced, the duration of the current and spontaneous radiation pulses decreased rapidly
within the range of laser emission conditions (Fig. 35). The relationship between the duration of the laser emission, current, and spontaneous radiation (from the upper level) pulses
was as shown in Fig. 35b. The intensity of the spontaneous radiation increased monotonically
with rising voltage and current (Fig. 36), and the dependence on the current was linear.

An investigation of the radiation emitted by the discharge in the visible part indicated
the presence (apart from CO) of the C_2 molecules. The maximum of the electronic C_2 bands
(mainly the Swan bands) corresponded to the trailing edge of the current pulse but the radiation started almost simultaneously with the beginning of the pulse. The intensity of the Swan
bands was comparable with the intensity of the CO bands and sometimes exceeded the latter.

Fig. 36. Dependences of the intensity of the
spontaneous emission of the 0−3 band in the
Ångstrom system of CO on the peak value of
the current (a) and on the voltage (b) for a
tube 8 mm in diameter.

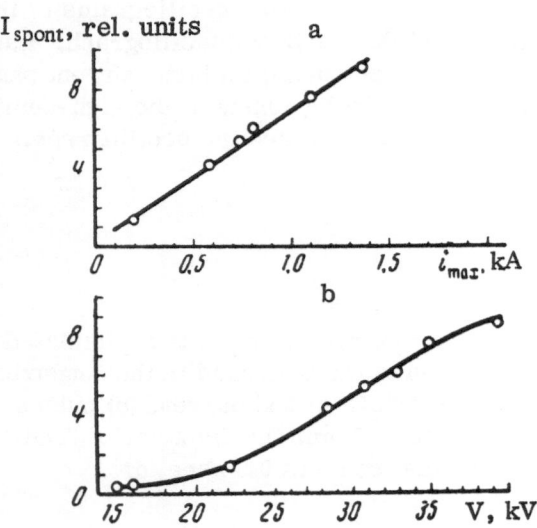

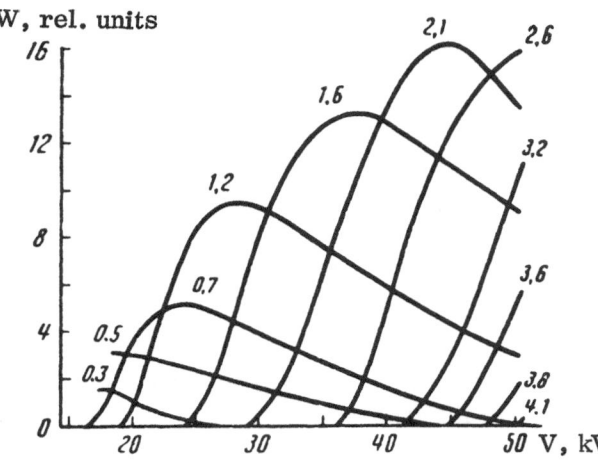

Fig. 37. Dependences of the average total laser emission power in the Ångstrom system of CO on the voltage across the working capacitor, obtained at various pressures of the working gas (given in Torrs alongside the curves).

These bands occasionally overlapped the CO bands which made it difficult to study the spontaneous emission spectrum of the CO molecule. Oscillograms of the spontaneous radiation were usually two-humped.

Investigations with Wide Tube [91]

For a number of reasons, we could expect interesting results when a large-diameter discharge tube was used. Consequently, we studied the laser emission in the Ångstrom bands in a tube whose diameter was 26 mm and which had the same cooled length as the narrow tube. The majority of the laser emission characteristics was determined when the temperature of the outer wall of the tube was 165°K.

Figure 37 shows the dependences of the total average laser emission power on the voltage across the working capacitor, obtained at various CO pressures. A comparison with the results for the narrow tube (Fig. 29a) demonstrated that the nature of the dependences of the laser emission power on the voltage and pressure was unaffected but the curves for the wide tube were displaced in the direction of higher voltages and they decreased more smoothly with rising voltage. These differences were manifested in a comparison of Figs. 38 and 29b, where the coordinates (V, N) were used to plot the points corresponding to the optimal case and to the limits of the range in which laser emission was observed. By analogy with the narrow tube, these points fitted well straight lines which converged at $V = V_0$ and the value of V_0 was practically the same for both tubes. Thus, the properties of the laser emission from the wide tube were also governed by the parameter γ. A comparison of the dependences of the

Fig. 38. Dependences of the range of existence of the laser emission in the Ångstrom system of CO on the voltage and gas density in a tube 26 mm in diameter (T = 165°K). The dashed line corresponds to the optimal conditions.

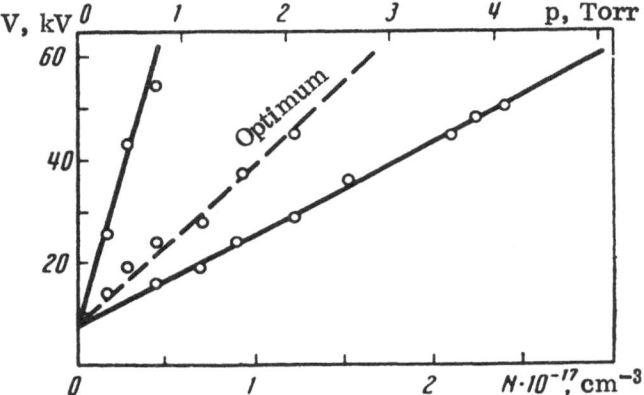

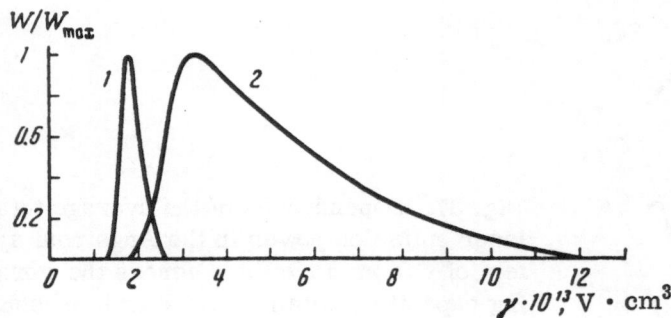

Fig. 39. Relative total laser emission power corresponding to the 0−4 band in the Ångstrom system of CO plotted as a function of the parameter γ: 1) tube of 8 mm diameter and with a discharge gap 111 cm long; 2) tube of 26 mm diameter and with a discharge gap 130 cm long.

relative power on the parameter γ is made in Fig. 39. Clearly, the range of laser radiation emitted from the 26-mm-diameter tube was considerably wider and γ_{opt} was displaced toward higher voltages. The values of γ_{opt} for the wide and narrow tubes were approximately in the ratio $(r_w/r_n)^{1/2}$, where r_w and r_n are the radii of the wide and narrow tubes, respectively. The dependences of the laser emission power on the voltage obtained for $\gamma = \gamma_{opt}$ were obtained also for the wide tube (Fig. 40). Throughout the investigated range of voltages, the power increased linearly with the voltage, whereas the narrow tube exhibited a maximum at 26 kV.

Measurements of the absolute value of the average total power were carried out for γ_{opt}, V = 30 kV, T = 165°K, and C = 0.012 μF. When the transmission of the exit mirror was 6%, which was optimal for the strongest (0−4) band, the output power was 160 W.

Our experiments demonstrated that a further cooling resulted in an additional increase in the gain and power. Therefore, we carried out some measurements with the tube cooled with liquid nitrogen. We found that in the wide tube the fall of the gas pressure in the discharge resulting from such cooling was much slower than in the narrow tube. This allowed us to investigate in detail the laser emission also when the tube was cooled with liquid nitrogen. However, considerable temperature and gas density gradients were established across the diameter of the wide tube. This was indicated by the observation that different bands were generated at different distances from the tube axis. Measurements of the power of the laser emission from the liquid-nitrogen-cooled tube were carried out as follows. The tube was cooled and filled with the active gas to a density exceeding the optimal value at the selected voltage. When the discharge was started, the CO pressure gradually fell, passing through its optimal value. We determined the maximum laser emission power corresponding to the optimal density of CO. The decomposition of CO in the cooled discharge resulted in deposition of the product on the tube windows and walls so that these had to be cleaned from time to time.

The first experiments on the liquid-nitrogen-cooled wide tube were carried out under conditions similar to those under which a peak power of 20 W was obtained from the narrow

Fig. 40. Dependence of the average total laser emission power in the Ångstrom bands of CO on the voltage ($\gamma = \gamma_{opt}$, T = 165°K, tube of 26 mm diameter).

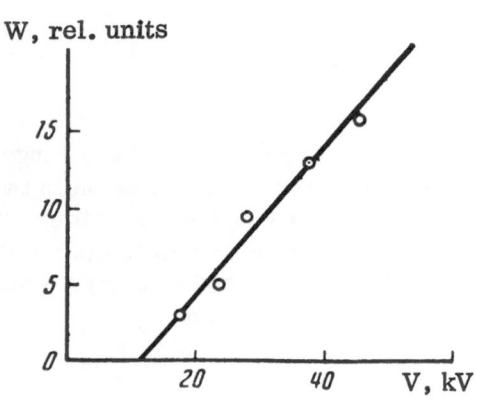

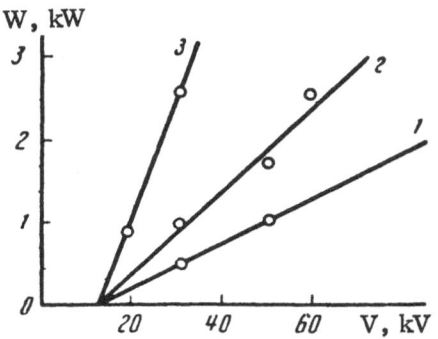

Fig. 41. Dependences of the total peak laser emission power in the Ångstrom bands of CO on the voltage applied to different working capacitors γ_{opt}, T = 80°K, tube of 26 mm diameter): (1) 0.012 μF; 2) 0.036 μF; 3) 0.5 μF.

tube (30 kV, 80°K, 0.012 μF). We then obtained the total peak power of 600 W. The transmission of the exit mirror was 6% and was clearly less than the optimal value. Thus, an increase in the discharge tube diameter from 8 to 26 mm increased the laser emission power by a factor of 30. However, larger capacitors and higher voltages could be used for discharges in the wide tube. Figure 41 shows the voltage dependences of the total peak laser emission power obtained for optimal gas densities using different working capacitors. The highest power was obtained for 0.036 μF at a pressure of 1.6 Torr when the repetition frequency was 2 Hz. This power reached 3 kW for 58 kV. This peak power was a record for this laser and it exceeded the one reported in [2] by almost three orders of magnitude. A similar power was obtained for 0.5 μF when the voltage was 30 kV and the pressure 0.3 Torr. In this case, the repetition frequency was governed by the charging time constant of the working capacitor and it was 0.65 Hz. The power obtained corresponded to a specific power of 6 W/cm^3, whereas for the narrow tube the corresponding value was 0.4 W/cm^3. Thus, not only the total power produced by the whole active medium was greater for the wide tube but also the specific power was larger and the latter increased by more than one order of magnitude. There was every reason to assume that a more careful optimization of the resonator and application of higher voltages to wider tubes should make it possible to achieve still higher output powers.

We also made an attempt to obtain continuous or quasicontinuous laser emission of the Ångstrom CO bands from the wide tube. In these experiments, the working capacitor was charged through a resistor and was connected parallel to the tube bypassing the discharger. Then, depending on the pressure and voltage, one could establish either a continuous or pulse (relaxation oscillations regime) discharge and the latter had a repetition frequency governed by the experimental conditions. Laser emission was not obtained from a continuously burning discharge but it was observed under pulse conditions even when the repetition frequency was high. The time intervals between the successive pulses were unstable and under certain conditions they amounted to 20–50 μsec, which corresponded to a repetition frequency of 20–50 kHz.

We also attempted to obtain quasicontinuous laser emission from apparatus in which gas could be excited by current pulses of variable duration ranging from 8 to 1000 μsec.* In this case, the voltage could be raised to 20 kV and the current amplitude could reach tens of amperes. The tube diameter was still 26 mm. The experiments were carried out under flowing-gas conditions with the outer wall of the discharge tube cooled to a temperature close to that of liquid nitrogen. Under these conditions, we observed laser emission of the Ångstrom band but again only in the form of pulses. The duration of these pulses was practically independent of the experimental conditions and it amounted to about 1 μsec. The laser action occurred at the beginning of the current pulse and could not be prolonged by extending the duration of the current pulse. The minimum amplitude of the current at which laser emission was still ob-

---

* The authors are grateful to S. V. Markova for an opportunity to carry out an investigation using apparatus generating current pulses of variable duration.

served was 15 A. The conditions favorable for laser emission depended on the pulse repetition frequency and were optimal when this frequency was ~100 Hz. The highest frequency under which regular laser emission pulses were observed was 2 kHz. The discharge in the tube was quenched when the repetition frequency was increased still further.

§2. Laser Emission Spectra

When we started the present investigation, the following six Ångstrom electronic bands were known to appear in laser emission [2, 58]:

Band $v' \to v''$ . . .	0—5	0—4	0—3	0—2	0—1	0—0
Band edge λ, Å .	6622	6080	5610	5198	4835	4511

TABLE 9. Rotational Structure of Laser Emission Bands in Ångstrom System of CO

λ, Å (in air)			Rotational transition
Our results	[2]	[58]	
0—5 band			
		6620.33 w	P4
6620.04		6620.03 w	P5
6619.29		—	P6
6618.18		6618.19	P7
6616.33		6616.34 w	Q2
6615.12		6615.12	Q3
6613.53	6613.5 0.02	6613.53	Q4
		6612.43 w	P10
6611.51	6611.5 0.2	6611.51	Q5
		6609.71 w	P11
6609.10	6609.1 0.4	6609.10	Q6
6606.30	6606.4 1	6606.30	Q7
6603.06	6603.1 0.7	6603.06	Q8
6599.46	6599.5 0.3	6599.48	Q9
	6595.5 0.06	6595.47	Q10
		6591.05	Q11
		6586.23	Q12
		6581.03	Q13
0—4 band			
6080.07		6080.07 ?	P4
6079.95		—	P3
6079.88		6079.87	P5
6079.35		6079.34 ?	P6,2
6078.50		6078.50	P7,1
6077.34		6077.31	Q1 P8
6076.66		6076.63 w	Q2
		6075.84	P9
6075.66		6075.66	Q3
6074.35	6074.2 0.03	6074.36	Q4
6072.72	6072.5 0.3	6072.70	Q5
6070.75	6070.5 0.6	6070.75	Q6
6068.48	6068.2 1	6068.48	Q7
6065.86	6065.7 0.8	6065.88	Q8
6062.93	6062.9 0.5	6062.96	Q9
		6056.19 w	Q11
		6052.32	Q12
		6048.16 w	Q13
		6043.69 w	Q14
0—3 band			
		5609.93 w	P5
		5609.58 w	P6
		5608.98 w	P7
		5608.11 w	P8
		5607.00 w	P9
		5606.93 ?	Q2

TABLE 9. Continued

λ, Å (in air)			Rotational transition	
Our results	[2]		[58]	

0−3 band

5606.17			5606,13 w	Q3	
			5605.63		P10
5605,15			5605.09	Q4	
5603.85	5603,8	0.02	5603.80	Q5	
5602.27	5602.5	0.3	5602,24	Q6	
5600.44	5600,4	1	5600.43	Q7	
5598.35	5598.3	1	5598.36	Q8	
	5596.0	1	5596.02	Q9	
	5593.4	0.5	5593.43	Q10	
	5590.6	0.3	5590.58	Q11	
			5587.48	Q12	
			5584,11	Q13	

0−2 band

5198,06			5198,07		P4.5
			5197.86		P6
			5197.43		P7
			5196.80		P8
			5195,95 w		P9
			5194.88		P10
			5194.72 w	Q3	
			5193.87 w	Q4	
			5193,63 w		P11
			5192.81	Q5	
			5191,54	Q6	
			5190.01	Q7	
			5188,43	Q8	
			5186.53	Q9	
			5184.42	Q10	
			5182.11	Q11	
			5179,59 w	Q12	

0−1 band

			4835,23		P5
			4835.03		P6
			4834.67		P7
			4831,62	Q4	
			4830.67 ?	Q5	
			4829,56	Q6	
			4828.29	Q7	
			4826.85	Q8	
			4825.24	Q9	
			4823,48	Q10	
			4821,55 ?	Q11	
			4819,47	Q12	

0−0 band

			4510.76 w		P6
			4508.21 w	Q3	
			4507.62	Q4	
			4506.90	Q5	
			4506.02	Q6	
			4505.01	Q7	
			4503.82	Q8	
			4502,48	Q9	

Note. The relative intensities are given in the second and third columns on the right of the wavelength: in the second column the intensity of the strongest line in each band is taken as unity; the letter "w" in the third column denotes weak lines. The results are given for the spectra obtained without forced cooling. The designation of the rotational transition includes J″ of the lower state. The symbol "?" is used for the lines whose identification is not yet certain.

TABLE 10. Rotational Structure of New Laser
Emission Band 0−6

λ_{meas}, Å (in air)	ν_{meas}, cm⁻¹ (in vacuum)	$\Delta\lambda = \nu_{meas} - \nu_{calc}$, cm⁻¹	Rotational transition
7248.001	13793.11	0.20	$P3,4$
7247.470	13794.12	0.20	$P2,5$
7246.353	13796.25	0.16	$P6$
7244.456	13799.86	0.24	$Q1$
7243.423	13801.83	0.22	$Q2$
7241.873	13804.78	0.18	$Q3$
7239.817	13808.70	0.14	$Q4$
7237.207	13813.68	0.12	$Q5$
7234.047	13819.72	0.18	$Q6$

Under our conditions [90], we observed four of them: 0−5, 0−4, 0−3, and 0−2. Table 9 gives the results of our measurements of the wavelengths of the rotational components of these bands and their interpretation. It should be noted that the lines listed in Table 9 were not all obtained under the same conditions. For comparison, we included in this table the wavelengths of the laser emission lines obtained by other authors. We found that the wavelengths agreed with the limits of the experimental error. Moreover, we listed in Table 10 the corresponding data for the 0−6 band in which we observed laser emission for the first time as a result of cooling of the gas [91]. We found that the published literature had no mention of this band in stimulated or spontaneous form.

Our experiments indicated that the distribution of the laser emission power between the bands depended largely on the selected mirrors and on the discharge parameters. Stimulated emission of a particular band could be achieved by using mirrors with certain reflection coefficients. When aluminized mirrors, characterized by approximately the same reflection for all the bands, were employed, we found that the highest gain corresponded to the 0−4 band. The gains of the 0−5 and 0−3 bands were considerably less and approximately equal. The gain of the 0−2 band was even less; the 0−6 band was the weakest.

The distribution of the laser emission power in the rotational structure was independent of the selected mirrors and was basically the same for all the laser emission bands. Only the Q and P branches appeared in laser emission. The power was concentrated in the Q branch (as in the spontaneous radiation case), whereas the P-branch lines were considerably weaker. The exception to this rule was the laser emission in the 0−2 band, in which the 5198 Å line belonging to the P branch was stronger than all the other lines in this band. Table 9 lists only this line because the other lines in the band, located on both sides of it, were much weaker and it was difficult to measure accurately their wavelengths. The P4,5 line was the last one in the edge of the 0−2 band. The R branch was not observed in the laser emission spectra.

Another common property of all the bands in the rotational laser emission spectra was the presence of a definite maximum in the power distributions within the branches. These distributions were similar to those observed in the spontaneous radiation spectra. Figure 42 shows the distributions of the laser emission power in the Q branch of the 0−4 band obtained at different temperatures of the discharge tube wall. We found that cooling shifted J_{max} in the direction of lower values and caused a strong rise of the power. Cooling of the gas from room temperature to 130°K increased the average laser emission power at J_{max} by a factor exceeding 50.

An analysis of the dependences in Fig. 42 yielded the following empirical relationship between the average laser emission power and the temperature of the gas of optimal density:

$$W(J_{max}) = Ce^{-\alpha T},$$

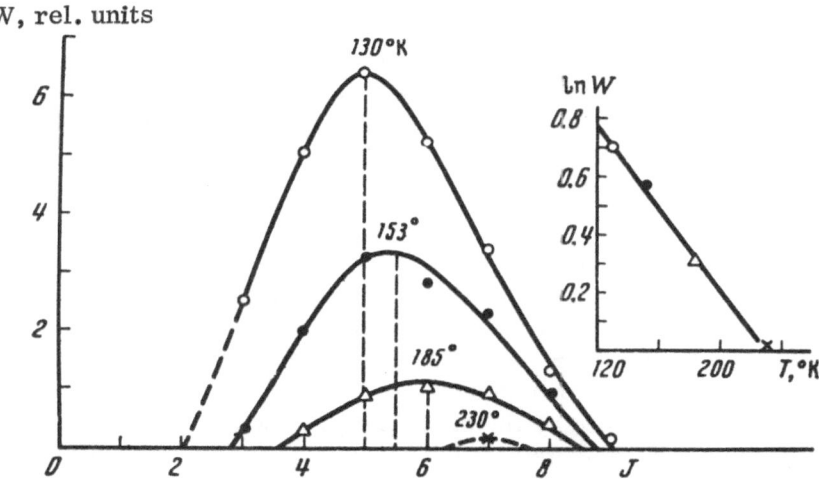

Fig. 42. Distributions of the laser emission power in the Q branch of the 0−4 band in the Ångstrom system of CO for various temperatures of the walls of a discharge tube of 8 mm diameter. The dependence ln W = f(T) is plotted for J_{max}.

where C and α are empirical constants. A linear approximation of the dependence ln W = f(T) (right-hand side of Fig. 42) was accurate to within 6%.

We shall consider in greater detail the new 0−6 laser emission band. It was first observed in a wide tube at 165°K when the pressure was 1 Torr, voltage was 26 kV, and transmission of the exit mirror in the laser emission region was 6%; the spectrum consisted of four Q-branch lines near 7240 Å. Cooling of the tube with liquid nitrogen increased the number of lines to nine and these included P-branch lines. A photograph of the spectrum of this band, obtained when the tube was cooled with liquid nitrogen, is shown in Fig. 43. The wavelengths of the new laser emission lines were measured using the DFS-13 spectrograph (dispersion 4 Å/mm) with absolute and relative errors of ~0.08 and 0.008 Å, respectively. Since this band had not been observed before and its rotational structure was not known, we carried out calculations using the following rotational constants:

$$\text{for } B^1\textstyle\sum^+ \quad v' = 0, \quad B_0' = 1.9475 \text{ cm}^{-1};$$
$$\text{for } A^1\Pi \quad v'' = 6, \quad B_6'' = 1.4667 \text{ cm}^{-1},$$

which were obtained from the formula $B_v = B_e - \alpha(v + \frac{1}{2})$ taking the relevant constants from [59].

A comparison of the experimental and calculated results demonstrated that laser emission was observed for the Q- and P-branch lines with values of J″ from 1 to 6. The measured wavelengths and frequencies of the laser emission lines and the identification of the transitions are given in Table 10. Some discrepancy was found between the calculated and measured values of the frequencies. The discrepancy increased from 0.29 to 0.90 cm⁻¹ when J″ in-

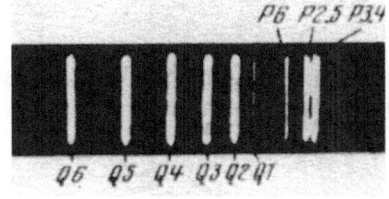

Fig. 43. Spectrogram of the laser emission of the newly discovered band 0−6 in the Ångstrom system of CO.

creased from 1 to 6. A more accurate agreement was obtained when we used the rotational constants $B_0' = 1.9536$ cm^{-1} and $B_0'' = 1.4556$ cm^{-1}. The difference $\Delta \nu$ between the measured frequencies and those calculated using the constants just given was tabulated in the third column of Table 10. A constant shift of ~0.2 cm^{-1} could be attributed to insufficient accuracy of the determination of the separation between the vibrational levels v' = 0 and v" = 6 ($T_{06} = 13{,}798.62$ cm^{-1}) used in the calculations.

§ 3. Discussion of Results

The experimental results reported in the present chapter demonstrated that the influence of the gas density and temperature and of the voltage across the tube on the laser emission of the Ångstrom CO band was basically the same as on the laser emission of the 1$^+$ and 2$^+$ nitrogen band systems. The properties of laser radiation emitted from the narrow and wide tubes were once again governed by the parameter γ. Hence, we concluded that the population inversion responsible for the Ångstrom bands was due to the interaction with electrons and its magnitude was governed by the temperature (or average energy) of the discharge electrons. A detailed investigation was made using tubes of two diameters. It demonstrated that the optimum values of γ_{opt} and the universal dependence of the relative laser emission power on the parameter γ were affected by the tube diameter: γ_{opt} increased with this diameter. Another important point was that the power of the laser emission from the wide tube obtained for $\gamma = \gamma_{opt}$ increased continuously with the voltage (and gas density), and with the working capacitance. This indicated that the power of the Ångstrom bands could be raised still further by increasing the voltage and capacitance and by optimization of the resonator. Moreover, we could expect that a further increase in the discharge tube diameter should also enhance the output power. Thus, our experiments demonstrated that a considerable laser emission power could be obtained in the Ångstrom CO bands and that there were good prospects for increasing this power to at least tens of kilowatts.

Cooling of the gas had a basically similar influence on the Ångstrom bands of carbon monoxide as on the nitrogen bands. Once again we observed a redistribution of the laser emission power in the rotational structure in such a way that the maximum shifted toward lower values of J, the gain increased strongly, and the output power became greater. However, in the case of CO, we observed a particularly rapid rise of the power, which could be due to the low initial room-temperature gain, which just exceeded the threshold.

The laser emission spectra in the Ångstrom band region were basically similar to the spectra reported elsewhere [2, 58]. Measurements of the wavelengths and their attributions were practically identical to those given by other workers and raised no doubts. The form of the laser emission spectrum as a whole was in good agreement with that expected on the basis of the Hönl–London factors of the relevant transition (see Chap. I). The observed relationship between the intensities of the various branches (predominance of the Q over the P branch and the absence of the R branch indicated that $N_v'/N_v'' > 1.5$ (Fig. 4).

The laser emission spectra also had some anomalies which did not fit the usual pattern. In particular, the power at 5198 Å in the 0−2 band was anomalously high and it exceeded considerably the power in other lines of this band. This anomaly could be explained [58] by the coincidence, within their widths, of two lines P4 and P5 so that the gains were combined and, consequently, the power increased considerably. A similar situation, although not so striking, was observed in the P6,2 and P7,1 lines in the 0−4 band.

When the gas was cooled, we observed for the first time a new (0−6) band which had not been found earlier even in the spontaneous radiation spectra.

This band was an example of how studies of the laser emission could refine the line positions and molecular constants.

We also found that the laser emission from CO had certain special characteristics associated with the asymmetry of the molecule. For example, when the gas was cooled, we found that, beginning from a certain temperature, the gas density in the discharge fell more or less rapidly. The fall was particularly strong in a narrow tube and, in this case, it started from 125°K. We could explain this behavior by postulating a considerable decomposition of CO in the discharge, resulting in the formation of various molecules (O_2, C_2, CO_2, O_3). This produced a dark deposit on the discharge tube walls and this deposit represented pure carbon and some of its compounds in the solid state. The decomposition process became much faster when the temperature was lowered, but it would still be too slow to explain the experimentally observed fall of the gas density. We could account for it only by assuming freezing of CO_2. The vapor pressure of CO_2 was 1 Torr at 138°K and it fell to 0.3 Torr at 120°K, so that CO_2 should freeze out at 125°K. The rapid fall of the pressure indicated that CO decomposed considerably in the discharge and, among other products, this produced CO_2 molecules.

On the other hand, the observation that the universal dependence on the parameter γ remained unchanged at all temperatures indicated that the composition of the gas in the discharge rapidly reached equilibrium, which was not shifted until CO_2 began to freeze out. The rate of the freeze-out process was clearly governed by the rate of diffusion to the walls. This was indicated by the observation that the rate of change of pressure as a result of cooling was much less (even down to 80°K) in the wide than in the narrow tube.

Only a small number of symmetric molecules suitable for our purposes exists in nature. Some of them (N_2, H_2, D_2) have already been used for stimulated emission. It follows that further search for new active molecular media must necessarily be extended to asymmetric molecules. Molecules are unavoidably decomposed in high-power electric discharges so that such asymmetric molecules may produce a considerable number of new molecules. The example of CO shows that even such a simple molecule may produce a large number of new products in a discharge. This should be borne in mind in the planning of experiments on asymmetric molecules. In particular, if a gas is cooled, some of the products of the discharge reactions will be deposited on the tube walls which will impede further cooling. Consequently, the important points to consider are the vapor pressures of the reaction products at a given temperature, rates of their diffusion, and dimensions of the discharge tube. The appearance of new molecules in a discharge may also play an additional role in the establishment and depletion of population inversion.

We shall begin our analysis of the population inversion mechanisms by considering some characteristics of the levels and transitions in the CO molecule which are relevant to the laser emission of the Ångstrom bands. Figure 44 shows schematically the distributions of the potential energy in the CO molecule. The vertical dashed lines represent the most probable region of excitation of molecular levels by direct electron impact from the ground state. The wave function of the lowest vibrational level of the ground electronic state is included in Fig. 44. The investigated laser emission of the Ångstrom bands is due to transitions between singlet electronic states $B^1\Sigma^+$ and $A^1\Pi$. The bands participating in the laser action are identified by arrows. Both laser (active) states have the same multiplicity as at the ground state $X^1\Sigma^+$ and are coupled to the latter by strong optical transitions (Hopfield−Birge and fourth positive systems, respectively). The upper state B has no radiative decay channels apart from that to the ground state and to the lower laser level; moreover, no cascade transitions terminating at this state are known. The lower laser state A decays radiatively only to the ground state but it also is the final state not only of the Ångstrom system but also of the Herzberg band system which begins from the $C^1\Sigma^+$ state. The last state decays also along two additional channels, one of which is a strong optical transition to the ground state (Hopfield−Birge system) and the other is a weak intercombination transition to the triplet state $a'^3\Sigma^+$ (Knauss system). The B and C states predissociate beginning from the vibrational levels v = 2 and v = 1, respec-

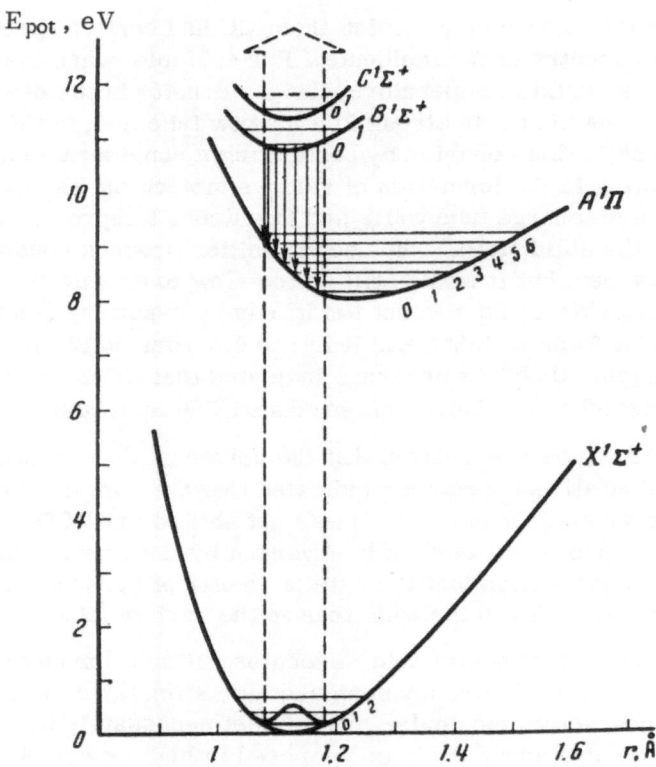

Fig. 44. Energy level scheme of the CO molecule.

tively. We must bear in mind that, under laser emission conditions, the radiation emitted as a result of transitions from the states A, B, and C to the lowest vibrational level of the ground state should be reabsorbed completely [90]. This alters considerably the real lifetimes of these states, particularly those of B and C. The lifetimes of the laser levels under total reabsorption and in its absence are listed in Table 11. This table includes the excitation cross sections of the laser levels by direct electron impact from the ground state of the molecule, taken from [74]. These cross sections are obtained in [74] on the assumption that the decay of the levels in the B state occurs only in transitions belonging to the Ångstrom system. However, it is clear from the results reported in [92] that the measurements in [74] were carried out under conditions such that the decay to the ground state was important. Consequently, we made the necessary corrections and then listed them in Table 11. A calculation of the life-

TABLE 11. Lifetimes and Excitation Cross
Sections of CO Levels

Electronic state	Vibrational level	Lifetime in the absence of reabsorption τ_0, nsec	Lifetime in the presence of reabsorption τ, nsec	Excitation cross section $\sigma_{max} \cdot 10^{18}$, cm^2
$B^1\Sigma^+$	$v'=0$	19,6	90.1	11.04
$A^1\Pi$	$v''=0$	10.9	12.7	1.70
	1	10.9	15.2	2.66
	2	10.5	15.2	3.02
	3	10.5	14.7	2.74
	4	10.4	12.7	2.06
	5	10.2	11.5	1.38
	6	—	—	—

times under total reabsorption conditions was made using the probabilities of spontaneous radiation from the active levels taken from [92-94] and making allowance for the relevant Franck—Condon factors taken from [92].

Since the laser emission of the Ångstrom bands is, on the whole, of the same nature as the laser emission of the 1^+ and 2^+ nitrogen systems, we have to assume that the principal population inversion mechanism is the excitation of the active levels by direct electron impact from the ground state of the molecule. This mechanism is supported by all the experimental results, including the appearance of the laser emission at the leading edge of the current pulses, linear rise of the stimulated radiation intensity with the current, and dependence of the power on V and N reducing to the universal dependence on the parameter γ. This mechanism is also supported by the laser emission spectra. The potential energy curve of the upper active state $B^1\Sigma^+$ is hardly shifted relative to the ground state. Then, it follows from the Franck—Condon principle that electron impact should excite effectively only the $v' = 0$ level of this state. All the observed laser emission bands begin from this level. However, in the case of this transition in CO, the Franck—Condon factors are not so favorable as for the 1^+ or 2^+ nitrogen systems. The excitation and emission channels of CO are not separated because the potential curves are not displaced. Therefore, all the laser emission bands terminate at the same levels of the state A which should be excited effectively by electron impact from the ground state. In this case, population inversion depends primarily on the ratio of the excitation cross sections of the vibrational levels. It is clear from Table 11 that these cross sections are such that we can qualitatively expect population inversion for the Ångstrom bands as a result of electron-impact excitation from the ground state of the molecule. Moreover, the transition probabilities suggest that this mechanism may also give rise to steady-state inversion of these transitions. However, only the pulse inversion has been observed experimentally.

In view of this situation, we calculated the gain for the Ångstrom bands of CO [95]. The calculation was carried out for steady-state (continuous) emission conditions on the assumption that the states A, B, and C were populated by direct electron impact from the ground state and by cascades from higher states. We assumed that the electron energy distribution was Maxwellian. These calculations demonstrated that the direct excitation by electrons should ensure population inversion of the bands for which laser emission has been observed. Moreover, steady-state inversion should be obtained for the Ångstrom band for any reasonable assumption about the temperature of electrons in the discharge. The gain should be sufficient to achieve continuous laser emission even at room temperature.

Thus, the calculations confirmed that the direct excitation by electrons should ensure population inversion and laser emission due to the transitions in question. On the other hand, some discrepancies between the calculated and experimental results were revealed. Particularly, the calculations predicted population inversion under steady-state conditions whereas laser emission was observed only under pulse conditions. Moreover, experiments indicated that the laser emission stopped much earlier than the emission of spontaneous radiation from the upper level, i.e., before the end of the pump pulse. Moreover, according to these calculations, the gain should increase continuously with increasing electron density n_e and electron temperature T_e, whereas we found experimentally that an optimum was observed in the dependence on the parameter γ (i.e., when n_e and T_e were increased), as demonstrated in Fig. 39; these features were demonstrated particularly strikingly in the narrow tubes.

It is interesting to compare the calculated and experimental distributions between the bands. The ratio of the maximum gains of the various bands calculated for average electron energies of 0.2 and 5 eV are plotted in Fig. 45. It is clear from these calculations [95] that, at electron temperatures expected in a discharge (2-5 eV), the greatest gain should be ex-

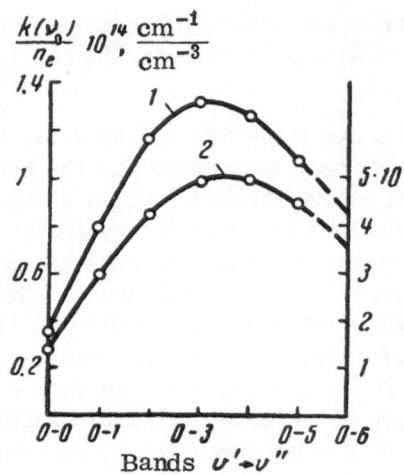

Fig. 45. Maximum gain in the Q branches of various bands in the Ångstrom system of CO at T = 315°K for average electron energies 5 eV (1) and 0.2 eV (2).

hibited by the 0−3 band. It is found experimentally that the 0−4 band has the highest gain. It is at present difficult to estimate the precision of our calculations and, therefore, it is impossible to say to what extent this discrepancy lies outside the range of the experimental error.

There is one further discrepancy between the calculated and experimental results. It follows from the calculations that the population of the lower level should be small compared with the population of the upper level. In this case, the optimal conditions are the same for all the bands because they all begin from the same upper level. However, it is found experimentally that the optimal conditions are very different for the various bands in question.

All this forces us to assume that there must be an additional process (ignored in the calculations) which fills the lower active level and which is responsible not only for the restriction of population inversion to pulse conditions but also for the difference between the optimal conditions for the various bands and possibly for the redistribution of the gain between the bands. Thus, the question of the additional process of the population of the lower level is closely related to the quenching of laser emission.

Since the duration of the laser emission increases with decreasing voltage (and current density) and since the specific laser emission power is higher for the wider tubes, it follows that the additional population of the lower level increases with the current density in the tube. Moreover, it is clear from the experimental data that the rate of pumping of the upper level also increases but inversion is no longer obtained.

Among processes which can populate additionally the lower level, it is worth considering primarily the radiative decay from the state C and superelastic collisions with electrons. It is suggested in [12] that the cascades from the C state are indeed responsible for the loss of inversion in the Ångstrom bands. This process was included directly in the calculations but it was found that the contribution of the cascade from the C state was small and the problem was still unsolved. Moreover, the excitation functions of the B and C states were known to be approximately the same [74] so that the relative role of the cascade should not change greatly with the discharge conditions.

Superelastic collisions with electrons may restrict considerably the attainable stimulated emission power and duration in pulse lasers [24, 96]. This is the process which limits the laser emission power due to transitions in the nitrogen molecule [24]. Since the electron densities in the experiments with CO were approximately of the same order of magnitude as in the experiments with nitrogen, this process could play a significant role. The increase in the role of the additional population of the lower level with rising electron density and the displacement (in the direction of longer delays) of the spontaneous radiation originating from the lower active level compared with the spontaneous radiation from the upper level could be regarded as arguments in support of this hypothesis.

Finally, we should also bear in mind the possibility of other processes. They include, for example, electron excitation of the higher vibrational levels of the ground state $X^1\Sigma^+$. A qualitative analysis based on the Franck−Condon principle (see the energy level scheme in Fig. 44) shows that the population of the higher vibrational levels of the ground state should reduce the electron pumping of the upper laser level and increase that of the lower laser level. Moreover, the reabsorption of the radiation originating from the $A^1\Pi$ state may be important. The existence of strong laser emission due to transitions between the vibrational levels of the ground state suggests that, under certain conditions, these levels may be heavily populated.

We should remember also possible processes resulting in the exchange of energy between the lower active level and levels of other electronic states in the molecule because of the existence of close resonances between them. For example, the energy defect between the $v = 4$ level of the A state and the $v = 14$ level of the $a'^3\Sigma^+$ state is only 23 cm^{-1}. We cannot exclude the possibility that the decomposition of CO in a discharge produces molecules or radicals which absorb in the laser transition range. The information available at present is insufficient to estimate the relative role of each of these suggested processes.

We may conclude that the laser emission in the Ångstrom band region is quasicontinuous because the duration of laser emission depends on the experimental conditions (it may follow the changes in the duration of the current and spontaneous radiation pulses and it may exceed considerably the lifetime of the upper active level) and because such emission is observed at high repetition frequencies (20-50 kHz). This means that continuous laser emission in these bands may be achieved under suitable conditions. This hope follows from the observation of the pulse laser emission at relatively low values of the current.

Thus, the available information shows that the main population inversion mechanism in the Ångstrom bands of the CO molecule is the excitation of the active levels by electrons directly from the ground state of the molecule. An analysis of these results shows also that there is an additional population process which applies to the lower level. The importance of this process increases with the electron density. This process limits the duration of laser emission and its power. The nature of this process cannot be determined without further studies. The available information suggests that the process is very likely to involve superelastic collisions with electrons.

CONCLUSIONS

The development of gas lasers has now reached a stage at which we know a considerable number of laser electronic transitions in diatomic molecules. The emission wavelengths of these lasers cover a wide spectral range from middle infrared to far ultraviolet. Record (for gas lasers) peak output powers have been obtained for them, high pulse repetition frequencies can be employed, and in some cases the efficiency is relatively high. This makes lasers utilizing electronic transitions in molecules very attractive for many applications in science and technology. However, for a long time the population inversion mechanism has not been firmly established because of lack of experimental data. Further successes in the development of these lasers can only be based on a fundamental (and best carried out by one team) study of their properties and characteristics, including optimal laser emission conditions and of physical properties responsible for population inversion.

We tried to carry out such a fundamental investigation covering all cases of practical importance in laser emission due to electronic transitions in molecules.

In the present chapter we shall compare the results for all the lasers which we investigated and consider characteristic features of the systems utilizing electronic transitions in molecules.

As pointed out in the Introduction, a distribution of population between a large number of sublevels in molecular-transition lasers results in a relatively low gain. Therefore, any possibility of increasing the gain, particularly in the case of pulse molecular systems, is of particular importance.

The present experimental and theoretical investigation has shown that one of the ways of increasing significantly the gain in an electronic-transition molecular system is cooling of the active gas. The influence of temperature on the laser emission properties was investigated for all laser systems. In all cases the influence of temperature was basically the same: Cooling of the gas resulted in a considerable increase in the gain and laser emission power, and it redistributed the power in the rotational structure of the band so that the maximum of the distribution shifted in the direction of lower values of J. Our preliminary results on the application of cooling to the $a^1\Pi_g - a'\ ^1\Sigma_u^-$ and $w^1\Delta_u - a^1\Pi_g$ infrared nitrogen band systems also indicated that the laser emission power increased considerably. It should be noted that in each specific case cooling to a given temperature increased the output power by an amount which varied from band to band: The increase was much greater for the weak bands. Thus, in the case of the second positive system of nitrogen for which fairly strong ultraviolet laser emission was observed at room temperature, cooling with liquid nitrogen increased the output power by just a factor of 1.5, whereas the increase for the weak laser emission in the Ångstrom bands of carbon monoxide was by more than two orders of magnitude. Cooling also enabled us to attain laser emission of new bands not observed before at room temperature (1−0 band in the 2^+ system of N_2, 0−6 band in the Ångstrom system of CO), i.e., we demonstrated experimentally the importance of cooling as a method in the search for new active media and the new laser emission lines.

The possibility of increasing greatly the gain by cooling the working gas was particularly important in the investigation of the laser emission of interest to us. It enabled us to obtain reliable information on various properties of the most important and typical laser emission due to electronic transitions in molecules, to identify conditions for achieving optimal output parameters, and to draw conclusions on the inversion population mechanisms. For example, the laser emission of the Ångstrom bands of the CO molecule had been hardly investigated because of its weak room-temperature gain.

We were able to identify some properties common to all the investigated laser systems. First of all, we noted the fact that for all systems the dependences of the laser emission and superradiance power on the voltage V across the working capacitor and on the gas density N could be represented by a universal function of the parameter $\gamma = (V - V_0)/N$. The same parameter described the lowest and highest values of the voltage and density at which laser emission appeared and disappeared, as well as the values at which the power reached its maximum value. The quantity V_0 in this parameter was found to be practically the same for all the investigated laser systems and it represented the voltage drop across external (located outside the active plasma) parts of the discharge circuit. In [57] a report was given of an investigation carried out under conditions similar to those employed in our case and it was shown that during the initial moments of the high-current phase of the discharge, i.e., when laser emission was observed from all our systems, distribution of the potential along the active plasma column in the tube could be regarded quite accurately as uniform. Thus, the parameter γ represented the voltage drop across the active plasma divided by the gas density, i.e., it was analogous to the ratio E/p used in plasma physics. This ratio [69] is known to represent the properties of the electron gas in the plasma, particularly the average electron energy.

In addition to detailed investigations of the laser emission in the 1^+ and 2^+ systems of nitrogen and in the Ångstrom system of CO, we investigated also — though in lesser detail — some other types of laser emission due to electronic transitions in molecules. They included transitions in the nitrogen systems $a^1\Pi_g - a'^1\Sigma_u^-$ and $w^1\Delta_u - a^1\Pi_g$ and in the hydrogen system $2s\sigma E^1\Sigma_g^+ - 2p\sigma B^1\Sigma_u^+$. Less attention was paid to these systems because the first two were not of great interest from the point of view of the physical mechanism of laser emission and the latter has already been investigated and its mechanism is known [5-7, 57]. We were interested in these systems from the point of view of the influence of the gas temperature and universality of the parameter γ. Laser emission due to transitions in N_2 has been discussed earlier and here we shall consider the laser emission from hydrogen. Pulse laser emission due to the $2s\sigma E^1\Sigma_g^+ - 2p\sigma B^1\Sigma_u^+$ electronic transition in the H_2 molecule and in its isotopes HD and D_2 was first discovered by Bazhulin et al. [5, 6]. They established, in particular, that the laser emission was due to the direct excitation of laser levels by electron impact from the ground state of the molecule. We found that cooling of the working gas increased considerably the output power and we established that the range of laser emission and the optimal conditions, considered as a function of the voltage and gas density, were governed by the parameter γ (Fig. 46). Thus, we found that in the case of laser emission due to electronic transitions in the H_2 molecule the physical mechanism was the same as for all the other systems investigated in detail.

The single-valued relationship between the properties of the investigated laser systems and the parameter γ, as well as the identical behavior of all the investigated systems when the working gas was cooled, indicated that the population inversion mechanism was the same in all of them. Moreover, the existence of the universal parameter γ showed that the processes of interaction of the active molecules with electrons played the dominant role in all these lasers.

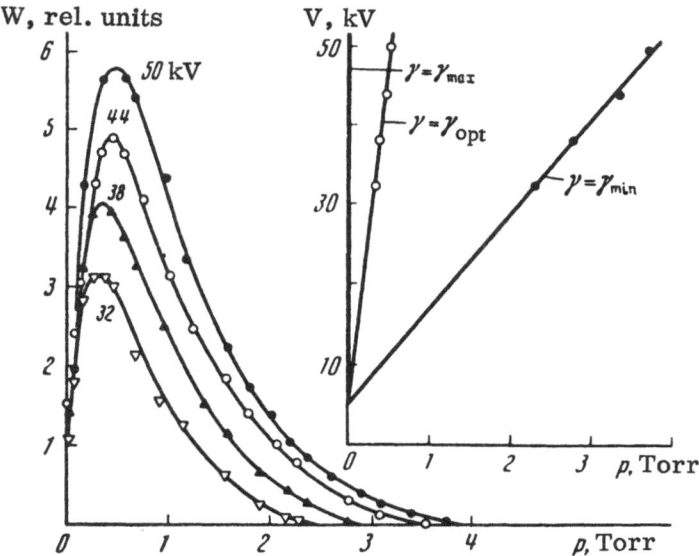

Fig. 46. Dependences of the average laser emission power due to the $2sE^1\Sigma_g^+ - 2pB^1\Sigma_u^+$ transition in the H_2 molecule on the gas pressure p obtained at different voltages V. The dependences on the right show the range of existence of laser emission plotted in the co-ordinates (V, N) and the conditions corresponding to maximum laser emission power (γ_{opt}) as a function of V and N.

The reported results enabled us to select the optimal conditions for the operation of lasers utilizing electronic transitions in molecules and to predict their ultimate parameters. Clearly, in all cases a change in the voltage across the tube should be accompanied by a corresponding change in the gas density so that the quantity $\gamma = \gamma_{opt}$ should be kept constant. For practically all the investigated system we observed a linear rise of the laser emission power with increasing voltage and gas density when the universal parameter had the value γ_{opt}. This revealed further opportunities for increasing the peak output power of all the investigated lasers by increasing the gas density and the electric field in the active plasma. Since it was inconvenient to increase further the working voltage, it would be preferable to employ sectioned tubes or use transverse discharges. It was demonstrated in [20, 27] that the use of a transverse discharge in a laser emitting in the 2^+ system of nitrogen made it possible to increase considerably the output power. Our estimates for the 1^+ system of nitrogen suggested that an output power of ~1 MW could be obtained. Our results indicated that under certain experimental conditions it should be possible to achieve the same results for other investigated lasers.

The increase in the laser emission power resulting from cooling enabled us not only to investigate the properties of the laser but also to improve the efficiency. The low gain of the molecular lasers affected particularly strongly the operation of pulse systems because in this case the inverted population existed for a finite and frequently very short time. In the case of infrared emission in the 1^+ nitrogen system we demonstrated that under normal conditions the laser gain and the rate of rise of the photon avalanche were such that saturation was not reached during the inversion lifetime. Consequently, the inverted population was largely unused and this reduced the peak output power and laser efficiency. Cooling and the resultant increase in the gain made it possible to approach more closely the saturation conditions and thus increase the peak output power and efficiency without introducing any additional energy into the active medium.

Our results demonstrated also that temperature was an important parameter for all the investigated laser systems utilizing electronic transitions in molecules. For example, in investigations of the dependences of the output power or efficiency on such parameters as the current, electron density, pulse repetition frequency, etc., whose increase could cause heating of the gas, the temperature of the gas had to be kept constant or its changes had to be recorded in order to obtain physically meaningful results. Otherwise, the recorded dependences would include the uncontrolled influence of changes in temperature which could be very considerable. For example, when the pulse repetition frequency was high (hundreds and thousands of hertz) in the 2^+ system of N_2, the fall in the laser emission power with rising frequency was mainly due to the heating of the gas [34, 35].

Temperature was particularly important when the active medium was a gas composed of asymmetric molecules (CO, NO, etc.). Since the nuclei of such molecules were not identical, a complex gas mixture was produced in the discharge and it consisted of the active substance as well as products of its decomposition. These products could have excitation and ionization potentials considerably lower than the corresponding potentials of the active gas. This would reduce the average electron energy in the discharge and, consequently, alter the population inversion and depletion processes. The relative concentrations of the active substance and of its decomposition products in a gas-discharge medium could depend strongly on the gas temperature. Thus, uncontrolled changes in the characteristics of the electron gas could result from changes in the gas temperature in lasers utilizing asymmetric molecules. This was an important factor because the interaction between molecules and electrons played the dominant role in lasers utilizing electronic transitions in molecules.

The similarity of the properties of the investigated lasers was not accidental. All the results obtained so far (those reported here and by other authors) indicated that the dominant pulse inversion mechanism in lasers utilizing electronic transitions in molecules (1^+ and 2^+ systems of N_2, Ångstrom system of CO, and $E^1\Sigma^+ - B^1\Sigma_u^+$ sytem of hydrogen) was the excitation of the active states by electron impact from the ground state of the molecule. The experimental observations reported in [10, 64, 65, 98-102] for all other lasers of this type were also in agreement with this mechanism. In all cases population inversion and distribution of the gain between the bands were subject to the Franck—Condon principle governing the excitation of optical transitions by electrons.

TABLE 12. Gas Lasers Utilizing Electronic Transitions in Molecules

Molecule, transition	Band	No. of lines in band	λ, Å	Ultimate efficiency, %	Experimental results			Reference
					P_{peak}	P_{peak}/V	τ_{las}, nsec	
N_2, $C^3\Pi_u - B^3\Pi_g$ (2$^+$)	0—1	11	3577	16	3 MW	40 kW/cm^3	4—7	14—16, 27—29
	0—0	43	3371	17				
	1—0	17	3159	18				
N_2, $B^3\Pi_g - A^3\Sigma_u^+$ (1$^+$)	0—1	7	12345	7	55 kW	0.3 kW/cm^3	100	17, 52—57, 88, 89, 97
	0—0	26	10490	8				
	3—3	13	9666	8				
	1—0	79	8900	9				
	2—1	65	8710	9				
	3—2	8	8530	9				
	2—0	41	7744	10				
	3—1	38	7617	10				
	4—2	61	7496	10				
N_2, $a^1\Pi_g - a'^1\Sigma_u^-$	0—0	2	8.2 μ	0.5				53, 98
	1—0	5	3.5 μ	1				
	2—1	6	3.3 μ	1				
N_2, $w^1\Delta_u - a^1\Pi_g$	0—0	12	3.6 μ	1.4				99
N_2, $B^3\Pi_g - {}^3\Delta_u$ (?)	—		5.2—6.3 μ					53
H_2, $E^1\Sigma_g^+ - B^1\Sigma_u^+$	0—2	2	15820	3	1,5 kW		100	57, 97, 99, 91, 100, 101
	0—1	2	13161	4				
	0—0	2	11222	4.5				
	1—0	3	8899	6.5				
	2—1	1	8350	6				
	2—0	1	7525					
H_2, $B^1\Sigma_u^+ - X^1\Sigma_g^+$ (Lyman)	5—12	3	1613	32	100 kW		1	64—66
	6—13	2	1609					
	4—11	2	1608	32.5				
	3—10	1	1596	33				
	7—13	1	1581					
H_2, $C^1\Pi_u - X^1\Sigma_g^+$ (Werner)	4—8	1	1240					
	3—7	1	1230					
	2—6	3	1219					
	1—5	1	1205					
	3—6	1	1189					
	2—5	3	1176					
	1—4	3	1160					
D_2, $E^1\Sigma_g^+ - B^1\Sigma_u^+$	1—0	3	9530	5				53, 57
	2—0	1	8278	6				
D_2, $B^1\Sigma_u^+ - X^1\Sigma_g^+$	—	—	1615—1520					65
HD, $E^1\Sigma_g^+ - B^1\Sigma_u^+$	1—0	1	9172	5,5				53, 57
HD, $B^1\Sigma_u^+ - X^1\Sigma_g^+$	—	—	1615—1520					65
CO, $B^1\Sigma^+ - A^1\Pi$ (Ångstrom)	0—6	9	7240	12	3 kW		100	2, 58, 97, 90, 91, 102
	0—5	32	6620	13				
	0—4	36	6080	14				
	0—3	22	5610	15.5				
	0—2	35	5198	17				
	0—1	12	4835	18				
	0—0	8	4511	19				
CO, $A^1\Pi - X^1\Sigma^+$	—	—	2000—1800					65
NO, $B'^2\Delta - C^2\Pi$	4—1	1	10215					10

The available evidence justifies treating gas lasers utilizing electronic transitions in molecules as a separate class. All known lasers belonging to this class share the same basic properties and the same population inversion mechanism. The characteristics of all the lasers in this class known at present are listed in Table 12. It is clear from this table that the number of molecules which can be utilized in such lasers is still small. It is limited to the simple and thoroughly investigated molecules N_2, H_2, CO, and NO. Nevertheless, laser emission has been observed in over 600 lines over a very wide spectral range from 8 μ to 1500 Å. Considerable peak output powers have been achieved.

Since the direct electron-excitation mechanism, responsible for the stimulated emission due to electronic transitions in molecules, is very general and can be applied to a large number of molecules, we may expect that further members will be added to this class of lasers. In particular, utilization of transitions from resonance electronic states to higher vibrational levels in the ground state should make it possible to extend the range further into vacuum ultraviolet and to build lasers with a high peak output power and high efficiency. It should be noted that extension in the direction of shorter wavelengths is made difficult by the rapid fall of the gain because it is proportional to λ^3. The increase in the gain with decreasing temperature may help in this respect.

There are also very good prospects of achieving continuous laser emission due to electronic transitions in molecules. The results of our investigations demonstrate, for example, that we can fully expect such continuous laser emission in the visible part of the spectrum due to transitions in the Ångstrom system of the CO molecule. Consequently, we can quite confidently expect lasers utilizing electronic transitions in molecule to find extensive applications in science and technology.

Obviously, further studies of lasers of this type should be carried out at low temperatures because cooling of the active gas helps greatly in the search for new molecular

TABLE 13. List of Potential Molecules for Laser Emission at Low Temperatures

Molecule	T (°C) corresponding to saturated vapor pressure of		Ionization potential, eV	Dissociation energy (eV) and process		Molecular weight, at. units
	1 Torr	10 Torr				
H_2	−263.3	−261.3	15.4	4.5		2
HD	—	−259.8	—	4.5		3
N_2	−226.1	−219.1	15.6	9.8		28
F_2	−223.0	−214.1	15.8	2.2		38
CO	−222.0	−215.0	14.1	9.1		28
O_2	−219.1	−210.6	12.2	5.1		32
F_2O	−196.1	−182.3	13.7	2.8	FO + O	54
CF_4	−184.6	−169.3	17.8	—		88
NF_3	Solid	−170.7	13.2	2.5	NF_2 + F	71
NO	−184.5	−178.2	9.5	6.5		30
O_3	−180.4	−163.2	>11.7	1.04	O_2 + O	48
HCl	−150.8	−135.6	12.8—13.7	4.43		36.5
N_2O	−143.4	−128.7	12.9	1.7 / 4.9	N_2 + O / NO + N	44
ClF	Solid	−139.0	—	2.6		54.5
HBr	−138.8	−121.8	—	3.8		81
CFN	−134.4	−118.5	—	—		45
CO_2	−134.3	−119.5	13.8	5.5	CO + O	44
H_2S	−134.3	−116.3	10.5	3.9	SH + H	34
SF_6	−132.7	−114.7	19.3	—		146
COS	−132.4	−113.3	—			60
NOF	−132.0	−114.3	—	2.4 / 6.3	NO + F / NF + O	49
HI	−123.3	−102.3	10.4	3.1		128
Cl_2	−118.0	−101.6	—	2.5		71
NH_3	−109.1	−91.9	10.3	4.5	NH_2 + H	17
Cl_2O	−98.5	−73.1	—	1.5	ClO + Cl	87
C_2N_2	−95.8	−76.8	13.6	—		42
SO_2	−95.5	−76.8	12.3	5.6	SO + O	64

laser emission and improves the characteristics of such emission. Therefore, it is interesting to consider the molecules which can be used at low temperatures. With this in mind we analyzed the properties of a large number of molecules from the point of view of laser emission at low temperatures. Since the population inversion mechanism discussed above should not be limited to diatomic molecules, we considered also polyatomic substances. Moreover, we could, in principle, expect inversion due to some products of decomposition of the original molecules in the discharge.

Table 13 lists the molecules in decreasing order of the possible degree of cooling. This list is limited to relatively simple molecules and then only those which can be cooled to about −100°C. Table 13 gives the temperature (in °C) at which the saturation vapor pressure lies between 1 and 10 torr. The table also gives the ionization potentials and dissociation energies of the molecules, as well as their molecular weights. In some cases the dissociation process is identified alongside the dissociation energy. It is clear from Table 13 that there is a fairly large number of molecules potentially suitable for laser emission due to electronic transitions at low temperatures. For example, these molecules include 12 diatomic substances on which experiments would be easiest to carry out. It should be noted that a deliberate search for molecules and electronic transitions is still quite difficult to carry out. This is due to the absence (in most cases) of such essential data as the excitation cross sections of the active levels, potential curves, and transition probabilities. The existence of a large number of molecules which can be cooled shows clearly that further extension of the class of gas lasers utilizing electronic transitions will be made experimentally on the basis of cooling of the active gas.

We can summarize our results as follows.

1. The hypothesis of the direct excitation of laser levels by electrons was used in a general theoretical analysis of the gain due to electronic transitions in diatomic molecules. Formulas were obtained for J_{max} corresponding to the maximum inversion in the rotational distribution. A specific calculation of the gain was made for the second positive system of N_2 and for the Ångstrom system of CO.

It was found that the gain corresponding to J_{max} increased strongly when the temperature of the working gas was lowered. For the rotational branches with $S_{J'J''} \sim J$ and high values of the vibrational inversion coefficient this rise was proportional to $1/T$.

2. A detailed experimental investigation was made of the spectral, energy, and time characteristics of the most important laser-emission electronic transitions in the 1^+ and 2^+ systems of N_2, Ångstrom system of CO, and $E^1\Sigma_g^+ - B^1\Sigma_u^+$ transition in H_2.

3. It was found that all these lasers systems exhibited dependences of the output power and duration of the laser emission pulses on the voltage V applied to the tube and the gas density N which could be reduced to the dependences on the parameters $\gamma = (V - V_0)/N$. The limits of the range of existence of laser emission and conditions corresponding to maximum power were characterized by definite values of γ.

4. It was found experimentally that, in agreement with calculations, cooling of the gas always increased strongly the gain and output power.

5. The laser emission and superradiance spectra of the 2^+ system of nitrogen were investigated for the first time. A detailed identification of the rotational transitions was given. Laser emission was observed in the $0-0$, $0-1$, and $1-0$ bands. The emission in the $1-0$ band appeared only when the working gas was cooled. The identification of the spectrum of the $0-0$ band was in agreement with the results reported by other workers.

6. A record peak laser emission power (55 kW) for the 1^+ nitrogen system was obtained by cooling the gas. Moreover, superradiance of 10 kW power was observed. Cooling increased

the output power and laser efficiency by a factor of 5. The physical efficiency of this system reached 0.2%, compared with the theoretical value of 1%.

The 1⁺ system was used to demonstrate experimentally the considerable role played by the rate of rise of the photon avalanches in pulse lasers.

Reversal of the intensity alternation in vibrational-rotational stimulated radiation spectra was reported for the first time.

7. Properties of the laser emission bands in the Ångstrom system of CO were investigated in detail for the first time. Cooling resulted in a record peak output power (for this system) amounting to 3 kW, which was almost three orders of magnitude higher than the reported value.

Cooling made it possible to observe for the first time the laser emission in the 0−6 band of this system. It had not been observed before even in spontaneous radiation.

Under certain experimental conditions we found that laser emission with a pulse repetition frequency of 20−50 kHz could be obtained. The emission in the region of the Ångstrom bands was quasicontinuous. In principle, it should be possible to achieve continuous laser emission in this system.

8. The hypothesis of population inversion by direct electron impact was confirmed experimentally for laser emission due to the $E^1\Sigma_g^+ - B^1\Sigma_u^+$ electronic transition in the H_2 molecule. Cooling gave a record peak power of 1.5 kW.

9. A total of about 180 new lines (about one-third known at present) were discovered and three new laser emission bands were found.

10. It was shown that the same population inversion mechanism applied to all the investigated laser systems. It was found to be based on the excitation of the active levels by direct electron impact from the ground state of the molecule. The distribution of the excitation and laser emission power between the bands was governed by the Franck−Condon principle.

The authors are very grateful to Professor N. N. Sobolev, Professor V. P. Tychinskii, V. A. Burmakin, and V. N. Ochkin for valuable discussions of the results obtained, and to I. N. Knyazev, S. V. Markova, M. D. Baranov, V. A. Yakovlev, and Z. É. Kun'kova for constructive discussions and help in the experiments.

APPENDIX

Summary Table of All Known Laser Emission Lines due to Electronic-Vibrational-Rotational Transitions in Molecules, Arranged in Order of Increasing Wavelength

The summary (Table 14) given below lists data on the following systems:

N_2: $B^3\Pi_g - A^3\Sigma_u^+$ ($B - A$, first positive system),
$C^3\Pi_u - B^3\Pi_g$ ($C - B$, second positive system),
$a^1\Pi_g - a^1\Sigma_u^-$ ($a - a'$), $w^1\Delta_u - a^1\Pi_g$ ($w - a$);

H_2: $2s\sigma E^1\Sigma_g^+ - 2p\sigma B^1\Sigma_u^+$ ($E - B$),
$C^1\Pi_u - X^1\Sigma_g^+$ ($C - X$, Werner system),
$B^1\Sigma_u^+ - X^1\Sigma_g^+$ ($B - X$, Lyman system);

D_2, HD: $2s\sigma E^1\sum_g^+ - 2p\sigma B^1\sum_u^+ (E - B)$;

CO: $B^1\sum^+ - A^1\Pi$ ($B - A$, Ångstrom system),

$A^1\Pi - X^1\sum^+$ ($A - X$, fourth positive system);

NO: $B'^2\Delta - C^2\Pi$ ($B' - C$).

The lines are arranged in order of increasing wavelength. The wavelengths obtained by different authors from laser emission spectrograms are listed: In the range below 2000 Å the values in vacuum are given, whereas those above 2000 Å are the values in air. In those cases where the original reference gave the wavelength above 2000 Å in vacuum [17, 52, 55], we converted the wavelengths to air. The wavelengths are given to the decimal place governed by the absolute error (~0.04 Å). For bands with closely distributed rotational lines the wavelengths are given to within the place governed by the relative error (~0.004 Å). In the latter case, the last figure is given as a subscript.

The wavelengths taken from [56] are calculated from tabulated energy levels of molecular nitrogen [68], the results of such calculations being either given in [68] or made in [56].

The standard notation system is used to identify the rotational transition. Table 14 gives the rotational branch (P, Q, R) and the rotational quantum number J" (of the lower active states), representing the total angular momentum of the molecule. The transitions in the first positive system of nitrogen (B — A, N_2) are the exception to this rule: In this case instead of J" the table gives K", which is the rotational quantum number of the lower active state representing the total angular momentum of the molecule after subtraction of the electronic spin momentum. The subscripts of P, Q, and R identify the electronic components of the multiplet splitting between which a given transition takes place. In the case of the symmetric molecules N_2, H_2, D_2, and HD the superscript "c" denotes the chief (strong) transition between the levels of the ortho modification and the prime is used for the transitions beginning from a specific component of the Λ splitting of the upper level of states with positive electronic orbital angular momentum Λ along the internuclear molecular axis (states Π, Δ, ...). For multiplet components with Ω = 0 and 2 of the $^3\Pi$ states of the N_2 molecule this is the Λ component with the lower energy, whereas for the $^3\Pi$ states (Ω = 1) this is the component with the higher energy.

The column headed "Reference" cites only that paper which gives the specific wavelength. The asterisk is used to identify the results obtained in the present investigation.

TABLE 14

Molecule	Electronic system	Vibrational band	λ_{meas}, Å	Rotational transition	Reference
H_2	$C-X$	1—4	1159.76	$R1^o$	104
			1161.36	$Q1^o$	104
			1166.17	$P3^o$	104
		2—5	1174.36	$R1^o$	104
			1175.86	$Q1^o$	104
			1180.5	$P3^o$	104
		3—6	1189.36	$Q1^o$	104
		1—5	1204.97	$R1^o$	104
			1206.68	$Q1^o$	104
		2—6	1217.34	$R1^o$	104
			1219.00	$Q1^o$	104
			1223.58	$P3^o$	104
		3—7	1230.04	$Q1^o$	104
		4—8	1239.56	$Q1^o$	104
	$B-X$	8—14	1567.3	$P3^o$	66
		2—9	1571.8	$P1^o$	66
		7—13	1577.1	$R1^o$	66
			1580.74	$P3^o$	64
		3—10	1591.5	$P1^o$	66
			1596.06	$P3^o$	64
		4—11	1604.48	$P1^o$	64
		6—13	1607.4	$P1^o$	66
		4—11	1608.39	$P3^o$	64
		6—13	1609.02	$P3^o$	64
		5—12	1610.33	$P1^o$	64
			1611.65	$P1^o$	64
			1613.18	$P3^o$	64
CO	$A-X$	2—6	1810.85	$Q\,(5-13),\ R\,(2-9)$	105
		2—7	1878.31	$Q\,(5-13),\ R\,(2-9)$	105
		3—8	1897.84	$Q\,(5-12),\ R\,(2-9)$	105
		2—8	1950.06	$Q\,(5-11),\ R\,(2-9)$	105
		3—9	1970.13	$Q\,(5-11)$	105
N_2	$C-B$	1—0	3157.56	$P_13^o,\ P_13$	*
			3157.78	$P_1'4^o,\ P_14$	*
				$P_22^o,\ P_2'2$	
			3157.98	$P_1'5,\ P_15^o$	*
			3158.03	$P_23,\ P_2'3^o$	*
			3158.16	$P_1'6^o,\ P_16$	*
			3158.27	$P_24^o,\ P_2'4$	*
			3158.32	$P_1'7,\ P_17^o$	*
			3158.44	$P_1'8^o,\ 8,\ P_1'15,15^o$	*
				$P_2'5^o,\ P_25$	
			3158.53	$P_1'9,\ 9^o,\ P_1'14^o,\ 14$	*
			3158.61	$P_1'10^o,\ 10,$	*
				$11^o,\ 11,\ 13^o,$	
				$P_1'13,\ P_1'12^o,\ 12,$	
				$P_2'6,\ 6^o,\ P_2'15^o,$	
				P_315^o	
			3158.70	$P_214,\ P_3'4^o$	*
			3158.74	$P_2'7^o,\ P_27$	*
			3158.83	$P_2'8,\ 8^o,\ P_2'13^o,$	*
				$P_2'14,\ 13$	
			3158.91	$P_2'9^o,\ 9,\ 10^o,\ 11,$	*
				$12^o,\ 10^o,\ P_2'11^o$	
				P_35^o	
			3159.00	$P_313^o,\ P_3'6^o$	*

TABLE 14. Continued

Molecule	Electronic system	Vibrational band	λ_{meas}, Å	Rotational transition	Reference
N_2	$C-B$	1—0	3159.11	$P_3'12^c$, P_27^c	∗
			3159.19	P_311^c, $P_3'8^c$	∗
				$P_3'10^c$, P_39^c	
		0—0	3364.90$_6$	R_17, $R_2'7^c$	∗
			3365.42$_5$	—	∗
			3365.48$_1$	$R_3'6^c$	∗
			3366.91$_5$	$R_3'4^c$	18
			3368.43$_5$	P_121, $P_3'20^c$	∗
			3369.25$_9$	P_11^c, $P_1'1$	∗
				$Q_2'1$, Q_21^c	
			3369.55$_5$	Q_32^c	∗
			3369.76$_9$	—	∗
			3369.84$_8$	P_13^c, P_217	∗
			3370.08$_5$	$P_1'4^c$, P_14	18
			3370.13$_7$	$P_1'16^c$	∗
			3370.17$_0$	P_116	∗
			3370.30$_5$	P_15^c	∗
			3370.32$_6$	$P_2'3^c$	∗
			3370.37$_2$	$P_1'15$	∗
			3370.44$_4$	$P_2'15^c$, P_315^c	17
			3370.47$_3$	$P_1'6^c$	∗
			3370.47$_9$	P_16	∗
			3370.53$_0$	$P_1'14^c$	∗
			3370.56$_0$	P_114	∗
			3370.56$_7$	P_24^c	∗
			3370.62$_8$	P_17^c	∗
			3370.66$_6$	P_214^c	∗
			3370.676$_1$	P_113^c	18
			3370.71$_7$	$P_3'14^c$	∗
			3370.72$_0$	$P_1'8^c$	∗
			3370.76$_2$	$P_1'12^c$, $P_2'5^c$	∗
			3370.80$_6$	P_19^c, $P_1'11$	∗
			3370.82$_0$	$P_1'10^c$	∗
			3370.84$_8$	P_111^c, P_110	∗
				$P_2'13^c$, P_213	
			3370.93$_1$	$P_2'6$, P_26^c	∗
			3370.94$_3$	P_313^c	∗
			3370.99$_6$	P_212^c, $P_3'4^c$	∗
			3371.04$_1$	$P_2'7^c$	∗
			3371.08$_9$	$P_2'11^c$, P_211	∗
			3371.12$_0$	$P_2'8$, P_28^c	∗
			3371.13$_6$	$P_2'10$, $P_3'12^c$	∗
			3371.14$_9$	P_210^c, $P_2'9^c$	∗
			3371.18$_5$	P_35^c	∗
			3371.27$_9$	P_311^c	∗
			3371.32$_0$	$P_2'6^c$	∗
			3371.37$_6$	$P_3'10^c$	∗
			3371 40$_1$	P_37^c	∗
			3371.41$_8$	—	19
			3371.43$_8$	$P_3'8^c$, P_39^c	∗

TABLE 14. Continued

Molecule	Electronic system	Vibrational band	λ_{meas}, Å	Rotational transition	Reference
N_2	$C—B$	0—1	3575.45$_9$	—	*
			3575.78$_9$	$P'_1 5$, $P_1 5^\circ$	*
			3576.09$_5$	$P_1 7$, $P_2 4^\circ$, $P'_2 4$	*
			3576.11$_1$	$P_1 7^\circ$	*
			3576.17$_3$	$P_1 11^\circ$	*
			3576.23$_5$	$P_1 10$, $P_1 9^\circ$	*
			3576.48$_6$	$P'_2 6$, $P_2 6^\circ$	*
			3576.61$_2$	$P'_2 9^\circ$, $P'_2 8$ $P_2 9$, $P_3 11^\circ$	*
			3576.61$_6$	$P_2 8^\circ$	*
			3576.89$_1$	$P'_3 6^\circ$	*
			3576.94$_9$	$P_3 7^\circ$, $P'_3 8^\circ$	*
CO	$B—A$	0—0	4502.48	$Q9$	58
			4503.82	$Q8$	58
			4505.01	$Q7$	58
			4506.02	$Q6$	58
			4506.90	$Q5$	53
			4507.62	$Q4$	58
			4508.21	$Q3$	58
			4510.76	$P6$	58
		0—1	4819.47	$Q12$	58
			4821.55	$Q11$	58
			4823.48	$Q10$	58
			4825.24	$Q9$	58
			4826.85	$Q8$	58
			4828.29	$Q7$	58
			4829.56	$Q6$	58
			4830.67	$Q5$	58
			4831.62	$Q4$	58
			4834.67	$P7$	58
			4835.03	$P6$	58
			4835.23	$P5$	58
		0—2	5179.59	$Q12$	58
			5180.77	—	97
			5182.11	$Q11$	58
			5184.42	$Q10$	58
			5185.76	—	97
			5186.53	$Q9$	58
			5187.74	—	97
			5188.06	—	97
			5188.43	$Q8$	58
			5188.98	—	97
			5189.39	—	97
			5190.01	$Q7$	58
			5191.54	$Q6$	58
			5192.81	$Q5$	58
			5193.63	$P11$	58
			5193.87	$Q4$	58
			5194.72	$Q3$	58
			5194.88	$P10$	58
			5195.42	$Q2$	97
			5195.95	$P9$	58
			5196.80	$P8$	58
			5197.43	$P7$ $(P2)$	58
			5197.86	$P6$ $(P3)$	58

TABLE 14. Continued

Molecule	Electronic system	Vibrational band	λ_{meas}, Å	Rotational transition	Reference
CO	B—A	0—2	5198.07	P5, 4	58
			5199.80	—	97
			5201.60	—	97
			5202.53	—	97
			5203.07	—	97
			5204.18	—	97
			5204.47	—	97
			5205.07	—	97
			5205.47	—	97
			5208.57	—	97
			5210.60	—	97
			5215.55	—	97
	B—A	0—3	5584.11	Q13	58
			5587.48	Q12	58
			5590.58	Q11	58
			5593.43	Q10	58
			5596.02	Q9	58
			5598.36	Q8	58
			5600.07	P13	97
			5600.43	Q7	58
			5601.20	R3	97
			5602.24	Q6	58
			5603.80	Q5	58
			5605.09	Q4	58
			5605.63	P10	58
			5606.13	Q3	58
			5606.93	Q2	58
			5607.00	P9	58
			5607.44	—	97
			5608.11	P8	58
			5608.98	P7	58
			5609.37	P2	97
			5609.58	P6	58
			5609.93	P5	58
		0—4	6043.69	Q14	58
			6048.16	Q13	58
			6048.59	R9	97
			6050.83	—	97
			6052.32	Q12	58
			6052.77	R8	97
			6056.19	Q11	58
			6056.61	R7	97
			6058.85	—	97
			6060.85	R6	97
			6062.15	—	97
			6062.96	Q9	58
			6063.22	R5	97
			6064.55	—	97
			6065.88	Q8	58
			6067.09	R4	97
			6068.48	Q7	58
			6070.75	Q6	58
			6072.70	Q5	58
			6074.36	Q4	58
			6075.66	Q3	58
			6075.84	P9	58
			6076.63	Q2	58

TABLE 14. Continued

Molecule	Electronic system	Vibrational band	λ_{meas}, Å	Rotational transition	Reference
CO	$B-A$	0—4	6077.31	$Q1$, $P8$	58
			6078.50	$P7$, 1	58
			6079.34	$P6$, 2	58
			6079.87	$P5$	58
			6079.95	$P3$	*
			6080.07	$P4$	58
			6080.77	—	97
			6083.04	—	97
			6085.02	—	97
			6088.87	—	97
			6091.14	—	97
			6093.11	—	97
		0—5	6581.03	$Q13$	58
			6583.90	—	97
			6586.23	$Q12$	58
			6586.64	—	97
			6589.13	—	97
			6591.05	$Q11$	58
			6592.72	—	97
			6593.03	—	97
			6595.15	—	97
			6595.47	$Q10$	58
			6598.11	—	97
			6599.48	$Q9$	58
			6601.74	—	97
			6603.06	$Q8$	58
			6606.30	$Q7$	58
			6609.10	$Q6$	58
			6609.71	$P11$	58
			6611.59	$Q5$	58
			6612.43	$P10$	58
			6613.53	$Q4$	58
			6615.12	$Q3$	58
			6615.94	—	97
			6616.34	$Q2$	58
			6617.99	—	97
			6618.19	$P7$	58
			6619.29	$P6$	*
			6619.38	($P6$)	97
			6620.03	$P5$	58
			6620.33	$P4$	58
			6622.47	—	97
			6625.00	—	97
			6628.57	—	97
		0—6	7234.05	$Q6$	*
			7237.21	$Q5$	*
			7239.82	$Q4$	*
			7241.87	$Q3$	*
			7243.42	$Q2$	*
			7244.46	$Q1$	*
			7246.35	$P6$	*
			7247.47	$P5$, 2	*
			7248.00	$P4$, 3	*
N_2	$B-A$	4—2	7465.62	—	*
			7469.31	—	*
			7471.17	—	*

TABLE 14. Continued

Molecule	Electronic system	Vibrational band	λ_{meas}, Å	Rotational transition	Reference
N_2	$B—A$	4—2	7471.68	$Q'_{23}11^c$	*
			7472.95	—	*
			7473.61	—	*
			7473.89	—	*
			7475.21	—	*
			7475.62	—	*
			7476.15	—	*
			7478.26	—	*
			7481.69	—	*
			7482.74	$Q'_1 11^c$	17
			7483.70	—	*
			7485.62	—	*
			7486.13	$P_{23}3^c$	*
			7487.41	$Q'_1 9^c$	17
			7488.24	$P_{23}5^c$	*
			7489.29	$P_{23}6$	*
			7489.77	$Q'_1 8$	*
			7489.83	$P_{23}7^c$	56
			7491.00	$Q'_{12}7^c$	*
			7491.72	$Q'_1 7^c$	*
			7492.69	—	*
			7492.96	$Q'_{12}6$	*
			7493.67	$Q'_1 6$	*
			7495.00	$Q'_{12}5^c$	*
			7495.67	$Q'_1 5^c$	*
			7496.65	—	*
			7496.86	$Q'_{12}4$	*
			7497.52	$Q'_1 4$	*
			7498.63	$Q'_{12}3^c$	*
			7499.36	$Q'_1 3^c$	*
			7500.09	$Q'_{23}7^c$	56
			7500.30	—	*
			7500.52	—	*
			7501.02	$Q'_1 2$	*
			7501.77	—	56
			7502.69	$Q'_1 1^c$	*
			7503.02	$P_{12}7^c$	*
			7503.69	$P_{12}5^c$	*
			7503.98	$P_{13}7^c$	*
			7504.47	$O'_{23}9^c$	56
			7504.56	$P_{12}5^c$	*
			7505.63	—	*
			7509.21	—	*
			7510.08	$O'_{12}3^c$	*
			7511.53	$O'_{12}4$	*
			7513.03	$O'_{12}5^c$	*
			7513.96	—	*
			7514.29	$O'_{12}6$	*
			7515.58	$O'_{12}7^c$	*
			7516.59	—	*
			7516.90	$O'_{12}8$	*
			7517.89	—	*

TABLE 14. Continued

Molecule	Electronic system	Vibrational band	λ_{meas}, Å	Rotational transition	Reference
N_2	$B-A$	4—2	7518.03	$O'_{12}9^c$	56
			7520.07	—	56
			7522.94	—	56
			7525.64	—	56
H_2	$E-B$	2—0	7525$\pm$0.5	$P2^c$	101
N_2	$B-A$	3—1	7572.25	$Q'_3 17^c$	55
			7574.40	—	56
			7581.05	$Q'_3 13^c$	52
			7584.19	$Q'_3 11^c$	52
			7586.47	$Q'_3 9^c$	52
			7587.74	$Q'_2 13^c$	55
			7589.89	$Q'_2 12$	55
			7591.80	$Q'_2 11^c$	55
			7592.03	$(Q'_2 11^c)$?	56
			7592.85	$R_1 8$	55
			7593.15	$Q'_1 15^c$	17
			7595.18	$Q'_2 9^c$	55
			7598.67	$Q'_1 13^c$	52
			7603.96	$Q'_1 11^c$	52
			7607.62	$P_{23} 3^c$	*
			7608.81	$Q'_1 9^c$	52
			7609.84	$P_{23} 5^c$	*
			7611.04	$Q'_1 8$	*
			7611.51	$P_{23} 7^c$	*
			7613.26	$Q'_1 7^c$	52
			7615.33	$Q'_1 6$	*
			7617.36	$Q'_1 5^c$	*
			7619.34	$Q'_1 4$	*
			7621.18	$Q'_1 3^c$	*
			7622.23	$O'_{23} 7^c$	*
			7622.92	$Q'_1 2$	*
			7622.96	$P_{12} 11^c$	52
			7624.21	$P_{12} 9^c$	17
			7624.65	$Q'_1 1^c$	*
			7625.14	$P_{12} 7^c$	*
			7625.66	$P_{12} 5^c$	*
			7626.01	$P_{13} 7^c$	*
			7626.04	$P_{12} 3^c$	56
			7626.23	$P_{12} 1^c$	56
			7626.73	$P_{13} 5^c$	*
			7626.81	$O'_{23} 9^c$	56
			7632.45	$O'_{12} 3^c$	*
			7633.97	$O'_{12} 4$	*
			7635.45	$O'_{12} 5^c$	*
			7636.84	$O'_{12} 6$	*
			7638.27	$O'_{12} 7^c$	*
			7639.57	$O'_{12} 8$	56
			7640.81	$O'_{12} 9^c$	*
			7641.95	$O'_{12} 10$	56

TABLE 14. Continued

Molecule	Electronic system	Vibrational band	λ_{meas}, Å	Rotational transition	Reference
N_2	$B—A$	2—0	7697.38	$Q'_3 17^c$	106
			7702.36	$Q'_3 15^c$	106
			7706.49	$Q'_3 13^c$	106
			7709.74	$Q'_3 11^c$	106
			7712.06	$Q'_3 9^c$ $(Q_1 17^c)$	106
			7713.35	$Q'_3 7^c$ $(Q_2 13^c)$	106
			7717.52	$Q'_2 11^c$	106
			7732.60	$Q'_1 10$	106
			7733.55	—	*
			7733.93	$P_{23} 3^c$	*
			7735.03	$Q'_1 9^c$	*
			7735.17	$P_{23} 4$	56
			7736.28	$P_{23} 5^c$	*
			7737.50	$Q'_1 8$	*
			7738.04	$P_{23} 7^c$	*
			7739.63	$Q'_1 7^c$	106
			7741.80	$Q'_1 6$	*
			7742.05	—	*
			7742.28	$P_1 18$	*
			7742.61	$Q'_{12} 5^c$	*
			7743.28	$Q'_{12} 4$	*
			7743.88	$Q'_1 5^c$ $(P_1 17^c)$	*
			7744.47		*
			7744.96	—	*
			7745.25	$Q'_{12} 4$	*
			7745.89	$Q'_1 4$	*
			7747.21	$Q'_{12} 3^c$	*
			7747.81	$Q'_1 3^c$	*
			7750.61	$P_1 11^c$	106
			7751.15	$P_{12} 9^c$	56
			7752.05	$P_{12} 7^c$	106
			7752.36	$P_{12} 6$	106
			7752.63	$P_{12} 5^c$	106
			7752.80	$P_{12} 4$	56
			7752.98	$P_{12} 3^c$, $P_{13} 7^c$	
			7753.74	$P_{13} 5^c$	*
			7754.10	$O_{23} 9^c$	56
			7754.70	$P_{13} 1^c$	106
			7759.62	$O'_{12} 3^c$	*
			7761.24	$O'_{12} 4$	56
			7762.82	$O'_{12} 5^c$	*
			7764.00	$O'_{12} 6$	*
			7765.77	$O'_{12} 7^c$	*
			7768.47	$O'_{12} 8$	56
			7773.16	$O'_{12} 9^c$	*
D_2	$E—B$	2—0	8277.52	$P 3^c$	53
H_2	$E—B$	2—1	8349.50	$P 2^c$	53
N_2	$B—A$	3—2	8518.42	$P_{23} 3^c$	56
			8520.95	$P_{23} 5^c$	56
			8524.88	$Q'_1 7^c$	56
			8530.38	$Q'_1 5^c$	56
			8535.38	$Q'_1 3^c$	56

V. M. KASLIN AND G. G. PETRASH

TABLE 14. Continued

Molecule	Electronic system	Vibrational band	λ_{means}, Å	Rotational transition	Reference
N_2	$B-A$	3—2	8539.91	$Q'_1 1^\circ$	56
			8540.82	$P_{12} 5^\circ$	56
			8541.45	$O'_{23} 9^\circ$	56
		2—1	8653.31	$R_3 7^\circ$	17
			8654.85	$Q'_3 15^\circ$	52
			8658.02	$Q'_3 14$	52
			8658.94	$S_{32} 1^\circ$	56
			8660.86	$Q'_3 13^\circ$	52
			8662.55	$Q'_2 15^\circ$	52
			8663.37	$Q'_3 12$	52
			8665.69	$Q'_3 11^\circ$	52
			8666.23	$Q'_1 17^\circ$	52
			8667.62	$Q'_3 10$	52
			8669.20	$Q'_3 9^\circ$	52
			8669.55	$Q'_2 13^\circ$	52
			8670.44	$Q'_3 8$	52
			8671.31	$Q'_3 7^\circ$	52
			8675.39	$Q'_1 15^\circ$	52
			8675.57	$Q'_2 11^\circ$	52
			8680.56	$Q'_2 9^\circ$	55
			8682.83	$R_1 7^\circ$	52
			8683.75	$Q'_1 13^\circ$	52
			8685.58	$Q'_{23} 7^\circ$	52
			8687.62	$Q'_1 12$	52
			8691.35	$Q'_1 11^\circ$	52
			8694.90	$Q'_1 10$	52
			8696.12	$P_{23} 2$	56
			8697.75	$P_{23} 3^\circ$	56
			8698.25	$Q'_1 9^\circ$	52
			8699.26	$P_{23} 4$	56
			8700.69	$P_{23} 5^\circ$	52
			8701.50	$Q'_1 8$	52
			8701.71	$P_{23} 6$	56
			8702.55	$P_{23} 7^\circ$	52
			8703.31	$P_{23} 11^\circ$	17
			8704.57	$Q'_1 7^\circ$	*
			8707.49	$Q'_1 6$	52
			8710.30	$Q'_1 5^\circ$	*
			8712.95	$Q'_1 4$	52
			8714.50	$P_1 13^\circ$	52
			8715.55	$Q'_1 3^\circ$	*
			8716.45	$P_{12} 11^\circ$	52
			8716.72	$Q'_{23} 7^\circ$	56
			8717.37	$P_1 11^\circ$	52
			8717.96	$Q'_1 2$	56
			8718.68	$P_{12} 9^\circ$	52
			8719.52	$P_{12} 8$	56
			8719.56	$P_1 9^\circ$	52
			8720.22	$P_{12} 7^c$	52
			8720.85	$P_{12} 6$	*

TABLE 14. Continued

Molecule	Electronic system	Vibrational band	λ_{meas}, Å	Rotational transition	Reference
N_2	$B—A$	2—1	8721.24	$P_{12}5^o$	*
			8722.01	$P_{13}3^o$	*
			8722.22	$P_2 5^o\ (P_{12}2)$	+
			8722.34	$P_{12}1^o$	52
			8722.37	$Q'_{23}9^o$	56
			8722.84	$P_{12}2$	*
			8726.33	$Q'_{12}1^o$	56
			8728.43	$Q'_{12}2$	56
			8730.45	$Q'_{12}3^o$	56
			8732.39	$Q'_{12}4$	56
			8734.25	$Q'_{12}5^o$	56
			8735.97	$Q'_{12}6$	56
			8737.64	$Q'_{12}7^o$	56
			8739.16	$Q'_{12}8$	56
			8740.56	$Q'_{12}9^o$	56
			8742.91	$Q'_{12}11^o$	56
		1—0	8833.74	$S'_{31}4$	52
			8841.27	$Q'_5 15^o$	52
			8842.4	$S'_{31}2$	*
			8844.18	$R_{21}9^o$	52
			8844.20	$R_{32}5^o$	56
			8844.55	$Q'_3 14$	52
			8845.42	$R_3 5^o$	56
			8847.58	$Q'_3 13^o$	52
			8849.19	$Q'_2 15^o$	52
			8850.24	$Q'_3 12$	52
			8852.61	$Q'_3 11^o$	52
			8853.00	$Q'_1 17^o,\ Q'_2 14$	52
			8854.63	$Q'_5 10$	52
			8856.28	$Q'_3 9^o$	52
			8856.54	$Q'_5 13^o$	52
			8857.56	$Q'_3 8$	52
			8857.86	$Q'_{32}5^o$	52
			8858.47	$Q'_3 7^o$	52
			8858.77	$R_1 9^o\ (Q'_3 4)$	52
			8859.09	$Q'_3 5^o$	56
			8859.77	$Q'_2 12$	52
			8861.54	$R_{21}5^o$	17
			8862.52	$Q'_1 15^o$	52
			8862.79	$Q'_2 11^o$	52
			8864.97	$Q'_1 14$	52
			8867.98	$Q'_2 9^o$	52
			8871.25	$Q'_1 13^o$	52
			8875.30	$Q'_1 12$	52
H_2	$E—B$	1—0	8876.13	$P4^o$	53
N_2	$B—A$	1—0	8879.14	$Q'_1 11^o$	52
			8882.83	$Q'_1 10$	52
H_2	$E—B$	1—0	8884.50	$P1$	95

TABLE 14. Continued

Molecule	Electronic system	Vibrational band	λ_{meas}, Å	Rotational transition	Reference
N_2	$B-A$	1—0	8884.55	$P_{23}2$	56
			8886.23	$P_{23}3^o$	56
			8886.41	Q'_19^o	52
			8887.78	$P_{23}4$	56
			8889.19	$P_{23}5^o$	52
			8889.70	Q'_18	52
			8890.26	$P_{23}6$	52
			8890.99	$P_29^o,\ 11^o,\ P_{23}13^o$	52
			8891.16	$P_{23}7^o$	52
			8892.17	$P_{23}9^o,\ 11^o$	52
			8892.97	Q'_17^o	52
			8895.97	Q'_16	53
H_2	$E-B$	1—0	8898.82	$P2^o$	53
N_2	$B-A$	1—0	8898.95	Q'_15^o	52
			8899.94	P_115^o	52
			8901.69	Q'_14	52
			8903.68	P_113^o	52
			8904.44	Q'_13^o	52
			8905.66	$P_{12}11^o$	52
			8906.12	$Q'_{23}7^o$	56
			8906.63	P_111^o	52
			8906.91	$P_{12}10$	56
			8907.01	Q'_12	56
			8907.93	$P_{12}9^o$	52
			8908.82	$P_{12}8$	56
			8908.88	P_19^o	52
			8909.55	$P_{12}7^o\ (Q'_11^o)$	52
			8910.11	$P_{12}6$	52
			8910.63	$P_{12}5^o$	52
			8911.01	$P_{12}4$	52
			8911.27	$P_{12}3^o$	52
			8912.17	$O'_{23}9^o$	56
			8915.83	$O'_{12}1^o$	56
			8918.06	$O'_{12}2$	56
			8920.20	$O'_{12}3^o$	56
			8922.27	$O'_{12}4$	56
			8924.24	$O'_{12}5^o$	56
			8926.12	$O'_{12}6$	56
			8927.89	$O'_{12}7^o$	56
			8929.53	$O'_{12}8$	56
			8931.04	$O'_{12}9^o$	56
			8932.41	$O'_{12}10$	56
			8933.61	$O'_{12}11^o$	56
HD	$E-B$	1—0	9172.01	$P2^o$	53
D_2	$E-B$	1—0	9441.56	$R1^o$	53
			9523.67	$P1^o$	53
			9530.05	$P3^o$	53
N_2	$B-A$	3—3	9650.89	$P_{23}3^o$	56
			9653.88	$P_{23}5^o$	56
			9655.66	$P_{23}7^o$	56
			9658.49	Q'_17^o	56
			9665.93	Q'_15^o	56

TABLE 14. Continued

Molecule	Electronic system	Vibrational band	λ_{meas}, Å	Rotational transition	Reference
N_2	$B—A$	3—3	9669.43	$Q_1'4$	56
			9672.71	$Q_1'3^c$	56
			9677.58	$P_{12}7^c$	56
			9679.43	$P_{12}5^c$	56
			9680.59	$P_{12}3^c$	56
			9690.92	$O_{12}'3^c$	56
			9695.18	$O_{12}'5^c$	56
			9698.80	$O_{12}'7^c$	56
NO	$B'—C$	4—1	10215 ± 20	$Q\ (\approx14^{1}/_{2})$	10
N_2	$B—A$	0—0	10435.88	$Q_1'15^c$	17
			10442.62	$Q_1'14$	17
			10449.03	$Q_1'13^c$	17
			10455.20	$Q_1'12$	17
			10461.17	$Q_1'11^c$	17
			10466.69	$Q_1'10$	17
			10471.95	$Q_1'9^c$	17
			10476.92	$Q_1'8$	17
			10479.57	$Q_{23}'7^c$	56
			10479.62	$P_{23}7^c$	17
			10481.74	$Q_1'7^c$	17
			10486.35	$Q_1'6$	17
			10490.64	$Q_1'5^c$	*
			10494.78	$Q_1'4$	*
			10498.72	$Q_1'3^c$	*
			10502.31	$P_{12}9^c$	17
			10505.12	$P_{12}7^c$	17
			10505.88	$Q_{11}'1^c$	56
			10507.13	$P_{12}5^c$	17
			10508.3_0	$P_{12}3^c$	17
			10520.81	$O_{12}'3^c$	56
			10522.6_0	$O_{13}'3^c$	17
			10526.2_2	$O_{12}'5^c$	17
			10530.9_3	$O_{12}'7^c$	17
			10534.7_1	$O_{12}'9^c$	17
H_2	$E—B$	0—0	11162.20	$P4^c$	53
			11222.50	$P2^c$	53
N_2	$B—A$	0—1	12302.61	$Q_1'11^c$	17
			12310.93	$Q_1'10$	17
			12318.82	$Q_1'9^c$	17
			12326.25	$Q_1'8$	17
			12333.34	$Q_1'7^c$	17
			12339.94	$Q_1'6$	17
			12346.31	$Q_1'5^c$	17
H_2	$E—B$	0—1	13056.62	$P4^c$	53
			13161.09	$P2^c$	53
		0—2	15641.93	$P4^c$	53
			15819.50	$P2^c$	53
N_2	$a—a'$	2—1	32937.24	$Q14^c$	17

TABLE 14. Continued

Molecule	Electronic system	Vibrational band	λ_{meas}, Å	Rotational transition	Reference
N_2	$a—a'$	2—1	33007.59	$Q12^c$	17
			33066.54	$Q10^c$	17
			33114.96	$Q8^c$	17
			33152.51	$Q6^c$	17
			33180.13	$Q4^c$	17
		1—0	34511.80	$Q12^c$	17
			34575.77	$Q10^c$	17
			34628.23	$Q8^c$	17
			34670.98	$Q6^c$	17
			34703.24	$Q4^c$	17
	$w—a$	0—0	36225.06	$R4^c$	97
			36251.53	$R3^c$	97
			36281.09	$R2^c$	97
			36421.39	$Q4^c$	97
			36437.24	$Q5^c$	97
			36456.21	$Q6^c$	97
			36478.35	$Q7^c$	97
			36503.81	$Q8^c$	97
			36532.40	$Q9^c$	97
			36564.49	$Q10^c$	97
			36599.54	$Q11^c$	97
			36638.34	$Q12^c$	97
			36679.85	$Q13^c$	97
			36725.17	$Q14^c$	97
			36773.37	$Q15^c$	97
	$a—a'$	0—0	81816.11	$Q8^c$	17
			82088.19	$Q6^c$	17

Literature Cited

1. L. E. S. Mathias and J. T. Parker, Appl. Phys. Lett., 3:16 (1963).
2. L. E. S. Mathias and J. T. Parker, Phys. Lett., 7:194 (1963).
3. H. G. Heard, Nature (Lond.), 200:667 (1963).
4. H. G. Heard, Bull. Am. Phys. Soc., 9:65 (1964).
5. P. A. Bazhulin, I. N. Knyazev, and G. G. Petrash, Zh. Eksp. Teor. Fiz., 47:1590 (1964).
6. P. A. Bazhulin, I. N. Knyazev, and G. G. Petrash, Zh. Eksp. Teor. Fiz., 49:16 (1965).
7. I. N. Knyazev, Zh. Prikl. Spektrosk., 3:510 (1965).
8. G. G. Petrash, Zh. Prikl. Spektrosk., 4:395 (1966).
9. P. A. Bazhulin, I. N. Knyazev, and G. G. Petrash, Zh. Eksp. Teor. Fiz., 48:975 (1965).
10. M. Huber, Phys. Lett., 12:102 (1964).
11. H. G. Cooper and P. K. Cheo, Bull. Am. Phys. Soc., 9:500 (1964).
12. P. K. Cheo and H. G. Cooper, Appl. Phys. Lett., 5:42 (1964).
13. H. G. Cooper and P. K. Cheo, Appl. Phys. Lett., 5:44 (1964).
14. V. M. Kaslin and G. G. Petrash, ZhETF Pis'ma Red., 3:88 (1966).
15. V. M. Kaslin and G. G. Petrash, Elektron. Tekh., Ser., 3, No. 2, 9 (1967).
16. V. M. Kaslin and G. G. Petrash, Zh. Eksp. Teor. Fiz., 54:1051 (1968); Preprint No. 132 [in Russian], Lebedev Physics Institute, Academy of Sciences of the USSR, Moscow (1967).
17. T. Kasuya and D. R. Lide, Jr., Appl. Opt., 6:69 (1967).
18. J. H. Parks, D. R. Rao, and A. Javan, Appl. Phys. Lett., 13:142 (1968).
19. M. Gallardo, C. A. Massone, and M. Garavaglia, Appl. Opt., 7:2418 (1968).
20. D. A. Leonard, Appl. Phys. Lett., 7:4 (1965).
21. E. T. Gerry and D. A. Leonard, Space/Aeronaut., 46:92 (1966).
22. J. D. Shipman and A. C. Kolb, Jr., IEEE J. Quantum Electron., QE-2, 298 (1966).
23. E. T. Gerry, Appl. Phys. Lett., 7:6 (1965).
24. A. W. Ali, A. C. Kolb, and A. D. Anderson, Appl. Opt., 6:2115 (1967).

25. A. W. Ali, Appl. Opt., 8:993 (1969).
26. Laser Focus, 3(5):39 (1967).
27. M. Geller, D. E. Altman, and T. A. DeTemple, Appl. Opt., 7:2238 (1968).
28. J. D. Shipman, Jr., Appl. Phys. Lett., 10:3 (1967).
29. E. L. Patterson, J. B. Gerardo, and A. W. Johnson, Appl. Phys. Lett., 21:293 (1972).
30. R. der Agobian, C. R. Acad. Sci. B, 263:1064 (1966).
31. L. Allen, D. G. C. Jones, and B. M. Sivaram, Phys. Lett. A, 25:280 (1967).
32. M. Garavaglia, M. Gallardo, and C. A. Massone, Phys. Lett. A, 28:787 (1969).
33. M. A. Kasymdzhanov, Vestn. Mosk. Univ. Fiz. Astron., 11:83 (1970).
34. V. A. Burmakin, A. A. Doroshkin, and G. G. Petrash, Elektron. Tekh., Ser. 1, No. 2, 142 (1970).
35. A. A. Doroshkin, Eletron. Tekh. Ser. 1, No. 9, 127 (1970).
36. A. Svedberg, L. Hogberg, Appl. Phys. Lett., 12:102 (1968).
37. J. Wilson, Appl. Phys. Lett., 8:159 (1966).
38. M. Geller, D. E. Altman, and T. A. DeTemple, J. Appl. Phys., 37:3639 (1966).
39. A. S. Nasibov, A. A. Isaev, V. M. Kaslin, and G. G. Petrash, Prib. Tekh. Eksp., No. 4, 232 (1967).
40. A. A. Podminogin, in: Gas-Discharge Devices (Proc. Third All-Union Conf. on Electronic Technology, Ryazan, 1970) [in Russian], No. 2 (18), Moscow (1970), p. 42.
41. K. G. Ericsson and L. R. Lidholt, Appl. Opt., 7:211 (1968).
42. D. A. Leonard, Laser Focus, 3(3):26 (1967).
43. M. Geller, D. E. Altman, and T. A. DeTemple, Appl. Phys. Lett., 11:221 (1967).
44. D. A. Leonard, Nature (Lond.), 216:142 (1967).
45. I. J. Eberstein, Rev. Sci. Instrum., 38:1665 (1967).
46. D. T. Phillips, Bull. Am. Phys. Soc., 13:1639 (1968).
47. H. P. Broida and S. C. Haydon, Appl. Phys. Lett., 16:142 (1970).
48. J. A. Myer, C. L. Johnson, E. Kierstead, R. D. Sharma, and I. Itzkan, Appl. Phys. Lett., 16:3 (1970).
49. J. A. Myer, I. Itzkan, and E. Kierstead, Nature (Lond.), 225:544 (1970).
50. G. Capelle and D. Phillips, Appl. Opt., 9:517 (1970).
51. L. R. Lidholt, IEEE J. Quantum Electron., QE-6:162 (1970).
52. I. N. Knyazev, Zh. Prikl. Spektrosk., 5:178 (1966); Preprint No. A-105 [in Russian], Lebedev Physics Institute, Academy of Sciences of the USSR, Moscow (1965).
53. R. A. McFarlane, in: Physics of Quantum Electronics (Proc. Intern. Conf., San Juan, Puerto Rico, 1965), publ. by McGraw-Hill, New York (1966), p. 655.
54. O. Andrade, M. Gallardo, and K. Bockasten, Appl. Opt., 6:2006 (1967).
55. C. A. Massone, M. Garavaglia, and M. Gallardo, IEEE J. Quantum Electron., QE-5:553 (1969).
56. L. N. Tunitskii and E. M. Cherkasov, Zh. Prikl. Spektrosk., 14:1004 (1971).
57. I. N. Knyazev, Thesis for Candidate's Degree [in Russian], Lebedev Physics Institute, Academy of Sciences of the USSR, Moscow (1968).
58. A. Henry, G. Arya, and L. Henry, C. R. Acad. Sci. (Paris), 261:1495 (1965).
59. G. Herzberg, Molecular Spectra and Molecular Structure, Vol. 1, Spectra of Diatomic Molecules, 2nd ed., Van Nostrand, New York (1950).
60. A. D. Sakharov, Izv. Akad. Nauk SSSR, Ser. Fiz., 12:372 (1948).
61. I. Kovács, Rotational Structure in the Spectra of Diatomic Molecules, A. Hilger, London (1969) and American Elsevier, New York (1970).
62. J. C. Polanyi, J. Chem. Phys., 34:347 (1961).
63. C. K. N. Patel, Phys. Rev., 141:71 (1966).
64. R. T. Hodgson, Phys. Rev. Lett., 25:494 (1970).
65. R. T. Hodgson, Bull. Am. Phys. Soc., 16:43 (1971).

66. R. W. Waynant, J. D. Shipman, Jr., R. C. Elton, and A. W. Ali, Appl. Phys. Lett., 17:383 (1970).
67. A. G. Fox and T. Li, Bell Syst. Tech. J., 40:453 (1961).
68. G. H. Dieke and D. F. Heath, Johns Hopkins University Spectroscopic Report, No. 17, Baltimore, Md. (1959).
69. S. C. Brown, Basic Data of Plasma Physics, MIT Press, Cambridge, Mass. (1959).
70. W. Benesch, J. T. Vanderslice, S. G. Tilford, and P. G. Wilkinson, Astrophys. J., 142:1227 (1965).
71. Y. Tanaka, J. Opt. Soc. Am., 45:663 (1955).
72. G. Chandraiah and G. G. Shepherd, Can. J. Phys., 46:221 (1968).
73. D. C. Cartwright, Phys. Rev. A, 2:1331 (1970).
74. V. V. Skubenich, Thesis for Candidate's Degree [in Russian], Uzhgorod (1969).
75. W. Benesch, J. T. Vanderslice, S. G. Tilford, and P. G. Wilkinson, Astrophys. J., 143:236 (1966).
76. J. D. Jobe, F. A. Sharpton, and R. M. St. John, J. Opt. Soc. Am., 57:106 (1967).
77. D. J. Burns, F. R. Simpson, and J. W. McConkey, J. Phys. B, 2:52 (1969).
78. W. Legler, Z. Phys., 173:169 (1963).
79. A. G. Engelhardt, A. V. Phelps, and C. G. Risk, Phys. Rev., 135:A1566 (1964).
80. P. N. Stanton and R. M. St. John, J. Opt. Soc. Am., 59:252 (1969).
81. J. W. McConkey and F. R. Simpson, J. Phys. B, 2:923 (1969).
82. M. Jeunehomme and A. B. F. Duncan, J. Chem. Phys., 41:1692 (1964).
83. N. P. Carleton and O. Oldenberg, J. Chem. Phys., 36:3460 (1962).
84. D. E. Shemansky and N. P. Carleton, J. Chem. Phys., 51:682 (1969).
85. D. E. Shemansky, J. Chem. Phys., 51:689 (1969).
86. N. I. Krindach and L. N. Tunitskii, Zh. Tekh. Fiz., 38:865 (1968).
87. K. V. Bol'shova and A. A. Shubin, Zh. Tekh. Fiz., 38:1090 (1968).
88. V. M. Kaslin, I. N. Knyazev, and G. G. Petrash, Kratk. Soobshch. Fiz., No. 1, 51 (1971).
89. V. M. Kaslin, I. N. Knyazev, and G. G. Petrash, Kvant. Elektron. (Mosc.), No. 5, 44 (1971).
90. M. D. Baranov, V. M. Kaslin, and G. G. Petrash, Zh. Eksp. Teor. Fiz., 57:375 (1969).
91. V. M. Kaslin, Z. É. Kun'kova, and G. G. Petrash, Kvant. Elektron. (Mosc.), No. 5, 101 (1972).
92. J. H. Moore, Jr., and D. W. Robinson, J. Chem. Phys., 48:4870 (1968).
93. J. E. Hesser and K. Dressler, Astrophys. J., 142:389 (1965).
94. J. E. Hesser, J. Chem. Phys., 48:2518 (1968).
95. V. M. Kaslin, Eletron. Tekh., Ser. 3, No. 1, 20 (1971).
96. A. A. Isaev, V. M. Kaslin, and G. G. Petrash, Preprint No. 81 [in Russian], Lebedev Physics Institute, Academy of Sciences of the USSR, Moscow (1970).
97. E. M. Cherkasov, Thesis for Candidate's Degree [in Russian], Lebedev Physics Institute, Academy of Sciences of the USSR, Moscow (1968).
98. R. A. McFarlane, Phys. Rev., 140:A1070 (1965).
99. R. A. McFarlane, Phys. Rev., 146:37 (1966).
100. K. Bockasten, T. Lundholm, and O. Andrade, J. Opt. Soc. Am., 56:1260 (1966).
101. R. M. Pixton and G. R. Fowles, IEEE J. Quantum Electron., QE-5:478 (1969).
102. V. M. Kaslin, G. G. Petrash, and V. A. Yakovlev, Kratk. Soobshch. Fiz., No. 7, 23 (1971).
103. R. W. Waynant, Phys. Rev. Lett., 28:533 (1972).
104. R. T. Hodgson and R. W. Dreyfus, Phys. Rev. Lett., 28:536 (1972).
105. R. T. Hodgson, J. Chem. Phys., 55:5378 (1971).
106. V. N. Ishchenko, V. N. Lisitsyn, and P. L. Chapovskii, Opt. Spektrosk., 33:366 (1972).